"In *Strategy-Driven Leadership*, Michael and Richard mark the route that every leader can take to assume the mantle of building that next generation for their business. Theirs is an insightful, practical and evidenced-based approach to developing the leaders our society needs now and in the future!"

Marshall Goldsmith
New York Times #1 bestselling author of *Triggers, Mojo,* and
What Got You Here Won't Get You There Thinkers 50 #1
Executive Coach for 10 years

"Too often there's a gap between what leadership development programs promise and what they actually deliver. Thank goodness, then, for Couch and Citrin. They begin with your organization's business strategy, build skills through actual experiences, and give readers an evidence-based process that delivers results."

Daniel H. Pink
New York Times, Wall Street Journal, Washington Post, Boston Globe, and
Amazon #1 bestselling author of *When, Drive, and To Sell Is Human*

"This book is a welcome sign of the times, addressing globalization, coaching, technology, and the integration of strategy. It is not, thankfully, a book of labels, and behavioral predictions, and cute descriptors of one's demeanor, which I find degrading and useless. Instead, it's a very thoughtful book for the mature leader to assess future conditions, needs, and new, potential leaders. It doesn't replace human judgment but rather augments it with solid methodology for assessment and selection. . . . Couch and Citrin have certainly gained my respect with this insightful work.

Alan Weiss, PhD
Author, *Million Dollar Consulting* and more than 60 other books
Hall of Fame Inductee: National Speakers Association
Lifetime Achievement Award: American Press Institute

"A must read for every CEO and CHRO frustrated by the elusive search for ready-now leaders who can actually execute a strategy and drive results."

Nancy Kukovich
President and CEO, AdelphoiUSA

"*Strategy-Driven Leadership* is a practical guide to defining, developing, and assessing the point in time leadership competencies an organization needs to produce desired business outcomes. It illustrates the connection between leadership behaviors and business performance and provides a framework for defining, assessing, and developing a path forward for closing gasps and improving business outcomes."

Sherry Neubert
Chief Information Officer, Goodyear Tire & Rubber Company

"Transformation is an urgent business requirement and *Strategy-Driven Leadership* is fundamental to that transformation. Michael and Richard have written a must read for those interested in adopting integrated strategies to accelerate the advancement of tomorrow's leaders."

James E. Taylor, Ph.D.
Chief Diversity, Inclusion, and Talent Management Officer
University of Pittsburgh Medical Center

"We often make the mistake of thinking that leaders are born, not made. But every leader has the opportunity to grow and improve—and *Strategy-Driven Leadership* provides the roadmap for organizations to develop their next generation of talent. Read this book, apply it, and watch how your company grows."

Dorie Clark
Author of *Reinventing You, Entrepreneurial You* and *Stand Out*
Executive Education Faculty, Duke University Fuqua School of Business

"*Strategy-Driven Leadership* is a must read for anyone who is interested in finding a research based practical approach to build the leadership muscle of a team, a division or an entire enterprise. Richard and Michael have developed a robust theoretical framework but perhaps even more importantly they go beyond just a framework and actually provide the tools to deploy this unique framework to grow the talent of the emerging leaders in a way that synchronizes with the strategic intent of your company."

Aradhna Malhotra Oliphant
President and CEO, Leadership Pittsburgh Inc.

"We have utilized aspects of *Strategy-Driven Leadership* and it has significantly enhanced our pipeline of leaders. You will find the approach and tools invaluable in growing your current and future leaders."

Stephanie Doliveira
VP of Human Resources, Sheetz Inc.

"Compelling, practical, and wise! *Strategy-Driven Leadership* fills a gap in our understanding of how to make leadership development really work."

Suzanne Bates
CEO, Bates Communications Inc.
Author of *All the Leader You Can Be, The Science of Achieving Extraordinary Executive Presence*

Strategy-Driven
Leadership

Strategy-Driven Leadership

The Playbook for Developing Your Next Generation of Leaders

Michael A. Couch and Richard S. Citrin

CRC Press
Taylor & Francis Group
Boca Raton London New York

CRC Press is an imprint of the
Taylor & Francis Group, an **Informa** business

First edition published in 2020
by Routledge/Productivity Press
52 Vanderbilt Avenue, 11th Floor New York, NY 10017
2 Park Square, Milton Park, Abingdon, Oxon OX14 4RN, UK

Routledge/Productivity Press is an imprint of Taylor & Francis Group, an Informa business

No claim to original U.S. Government works

Printed on acid-free paper

International Standard Book Number-13: 978-0-367-33226-6 (Hardback)
International Standard Book Number-13: 978-0-429-31909-9 (eBook)

Library of Congress Cataloging-in-Publication Data

Names: Couch, Michael A., author. | Citrin, Richard, author.
Title: Strategy-driven leadership : the playbook for developing your next generation of leaders / Michael A. Couch and Richard S. Citrin.
Description: New York, NY : Routledge, Taylor & Francis Group, 2020. | Includes bibliographical references and index.
Identifiers: LCCN 2019035543 (print) | LCCN 2019035544 (ebook) | ISBN 9780367332266 (hardback) | ISBN 9780429319099 (ebook)
Subjects: LCSH: Leadership.
Classification: LCC HD57.7 .C685 2020 (print) | LCC HD57.7 (ebook) | DDC 658.4/092—dc23
LC record available at https://lccn.loc.gov/2019035543; LC ebook record available at https://lccn.loc.gov/2019035544

Visit the Taylor & Francis Web site at
www.taylorandfrancis.com

To Dean, Roberta, Amy and Taylor

Couch who have all contributed

significantly to my personal development

Michael

To my wife, Sheila Collins, who

has always believed in me.

Richard

Contents

Acknowledgments xi
Authors xiii
Foreword xv

Chapter 1 Strategy-Driven Leadership Development 1

    A   A Leadership Development Fairy Tale 1
    B   The Status of Leadership Development Today 3
    C   How to Use This Book 10

Chapter 2 The Foundations of Strategy-Driven Leadership
Development 13

Chapter 3 Organizational Demand: What Do We Need
for Success? 21

    A   Strategy-Driven Competencies 24
    B   Identifying and Validating Strategy-Driven
          Leadership Competencies 25

Chapter 4 Assessing Organization and Leadership
Capability 35

    A   What to Look For: Growth Potential 37
    B   What to Look For: Performance Effectiveness
          Over Time 43
    C   How to Find It: The Best Assessment Tool
          or Method 45
    D   Our Answer: Talent Reviews 48
    E   How Talent Reviews Work 50
    F   The Development Strategies Matrix or the
          Nine Box on Steroids 58
    G   Development Strategies 61

Chapter 5 The Solution: Intentional Development 67

 A Why This Works: Learning Theory and
  Intentional Development...67
 B Intentional Leadership Development Essentials........71

Chapter 6 The Intentional Development Process............................ 87

 A Introduction 87
 B The Intentional Development "It" Formula:
  Frame It, See It, Own It, Connect It............................ 88
 C Frame It: Establish and Connect the
  Business Case 90
 D See It: Identify Essential Skills....................................94
 E Own It: Build the Skills..105
 F Connect It: Leverage New Skills.................................118

Chapter 7 Coaching Reimagined: Building Intentional
 Coaching.. 123

 A What Is Intentional Leadership Coaching? 126
 B Keys to Effective Intentional Coaching.....................127
 C Leaders as Coaches: Reimagining Your
  Approach..133

Chapter 8 Gauging Impact: Evaluating Intentional
 Development and Talent Analytics............................... 141

Chapter 9 The Strategy-Driven Leadership Development
 Journey.. 153

 A SDLD Maturity: It's a Journey, Not a
  One-and-Done 155
 B Strategy-Driven Leadership Development
  Derailers...157

Chapter 10 Conclusion 167

References.. 173
Index... 179

Acknowledgments

As you will see, our journey to the formation of Strategy-Driven Leadership Development and Intentional Leadership Development was a long and circuitous one, a journey that included the inspiration and insights from a number of colleagues, clients, and thought leaders. We have acknowledged many of them throughout the book. We would be remiss if we did not recognize others.

Our new paradigm for leadership development has benefited from the continued development and honing of competency-based talent management. We want to honor the contributions of our brilliant friend and colleague, Jane Schenck. Michael worked with Jane on several projects when he was in corporate roles. Jane has been our competency guru when it comes to our work with our consulting clients. Jane knows more about competencies and their application to talent challenges than anyone we have ever met. She was an early adopter as an associate of Lombardo and Eichinger and has stayed on top of competency technology ever since. Jane's influence on our knowledge is wide-ranging and is therefore hard to pinpoint. Jane's hands (and mind) are all over anything we have to offer in this book on competency-based talent management.

Our ideas are built on both the research efforts of practitioners who have gone before us and our clients who have allowed us into their workspace to contribute and learn from their challenges and successes. We hope we have properly acknowledged the assistance of the professionals whose research preceded our efforts. We are especially grateful to two of our clients, Sheetz Inc. and Adelphoi USA, who were early adopters of our strategy-driven approach to leadership development.

Our work would not have been successful if we did not have a strong business coach and adviser, and Alan Weiss has served in that role, especially for Richard. Alan's advice and urging about the importance of developing and sharing intellectual property for the advancement of the field as well as for building successful businesses has been invaluable.

Authors

Michael A. Couch has made a career out of improving the effectiveness of organizations, as an internal consultant, a business executive, and now in his own consulting practice. Starting from a graduate degree in Industrial and Organizational Psychology, he has more than 30 years' experience working on a wide range of organization challenges at the team, facility, division, and corporate levels.

In additional to his talent management and organization effectiveness background, Michael has the unique experience of being the Managing Director for a Strategic Business Unit as well as heading up Manufacturing, including Customer Service, Materials Management, Production Planning, Inventory Management and Logistics. He has over 25 years' experience with continuous improvement initiatives, from Quality Circles to Lean Six Sigma, and has led company-wide Lean Manufacturing implementations. He is a Certified Six Sigma Green Belt and Champion.

His expert facilitation skills have been applied to strategic planning, change management, team development, process improvement, project planning, and problem-solving situations. Michael is certified in Compression Planning® by The Compression Planning Institute. He has designed and facilitated sessions of all types—from small teams to groups over 200, from short one-to-two hour sessions to multi-day events—in a wide variety of industries.

Since its founding in 2007, Michael A. Couch & Associates Inc. has served more than 50 clients in a wide variety of industries, both domestic and global, and for-profit and non-profit.

Michael is a certified Human Capital Strategist by the Human Capital Institute and a Senior Professional in HR by the Society for Human Resources Management.

You can learn more about Michael at his website: www.mcassociatesinc.com.

 **Richard S. Citrin** bridges the gap between business and psychology. He draws on his deep experience as a health care entrepreneur, clinician, corporate executive, and leadership coach to present his insights and ideas in a practical and powerful manner that creates dramatic change and success for his clients.

His unique perspective on leadership, talent development, and how we learn and grow from challenges is based on the idea that by using our strengths, talents, and skills, and by aligning them in a purposeful and powerful manner, real change occurs.

Richard has published more than 30 professional and popular articles, is a frequent speaker and his ideas have appeared on the pages of *Forbes*, *Fortune*, and the *New York Times*. Richard's two prior books, *The Resilience Advantage* and *How to Get a Job in Dallas/Fort Worth*, provided his readers with information that helped them transform their professional and personal lives.

Foreword

There are 36 species in the cat family. They range from a housepet that weighs a few pounds if svelte and perhaps 20 if obese, to a Siberian tiger that hits the scales at more than 1,000 pounds. That's quite a divergence.

However, they're all cats. They look quite similar, although in different sizes. They groom in the same fashion, use their teeth and claws in similar fashion, and even purr alike, albeit lions can be heard purring up to five miles away, which is preferable to hearing them roar close by, one supposes.

We tend to regard leaders as cats. They are different in size and shape and gender, but don't all humans share common human traits, analogous to shared feline traits?

That, however, is a grossly incorrect view.

Apart from a common membership in the human race, there was and is very little in common among leaders such as Steve Jobs, Jack Welch, Oprah Winfrey, Dwight Eisenhower, Barack Obama, Sitting Bull, Roger Ailes, Bill Belichick, Francis Hasselbein, Nancy Pelosi, and all the others.

That's because cats are determined by DNA and "programmed" by that chemistry. Beyond DNA, people are sentient and learn and grow and can organize their own metamorphosis. As an example, Jack Welch went from "neutron Jack" to a much more sensitive and comprehensive executive as he led GE to its greatest heights, not since reached again.

Leadership is a learnable skill. I've always seen leaders, over 30 years of consulting globally, as "made" and not "born," despite culture and environment. The challenge for organizations and existing leadership is to identify those individuals with the highest capacity for learning and the greatest match with organizational strategy.

Couch and Citrin herein refer to it as "strategy-driven." This is a novel and highly useful approach.

This book looks to the next generation of leaders and how to find them and prepare them. Unlike a cat (or a tulip or a pelican), leaders are not successful merely by dint of continuing the species. They are successful by creating the next generation, which is adept at becoming part of, and then leading, the business case: the mission, or raison d'êtreof the enterprise.

The authors identify essential skills, talents, and behaviors that are essential both in general and for specific environments. They provide assessment tools based in learning theory and valid research. The have married the art and the science.

The book is a welcome sign of the times, addressing globalization, coaching, technology, and the integration of strategy. It is not, thankfully, a book of labels, and behavioral predictions, and cute descriptors of one's demeanor, which I find degrading and useless.

Instead, it's a very thoughtful book for the mature leader to assess future conditions, needs, and new, potential leaders. It doesn't replace human judgment but rather augments it with solid methodology for assessment and selection.

And in this manner it does share attributes with cats: the authors are agile, fast, and quite independent in their approach. Cats don't seek affection, the simply gain respect through their independence.

That's how I see Couch and Citrin, and they've certainly gained my respect with this insightful work.

Alan Weiss, PhD
Author, *Million Dollar Consulting* and more than 60 other books
Fellow: Institute of Management Consultants
Hall of Fame Inductee: National Speakers Association
Lifetime Achievement Award: American Press Institute

1

Strategy-Driven Leadership Development

A. A LEADERSHIP DEVELOPMENT FAIRY TALE

Once upon a time, leaders from a global creator and purveyor of goods converged on a resort for what was supposed to be three days of leadership development. The program included motivational speeches, leadership development exercises, and time together to bond over golf, tennis, and visiting local antique shops in the afternoons. The team running the event had spent the last year planning and organizing the conclave and believed they created a first-class program at a first-class property at a first-class price.

Much to their excitement, a new CEO had recently joined the firm, and while he was in favor of the event, he also had some direct and perhaps seemingly simple questions that he wanted answered. The first two were "How did all this get started?" and "Why are we doing this?" As the story went, the idea for the annual event had come from a previous executive who felt that the business's leaders needed to have leadership development training. The tradition had continued on for many years. When the participants were asked if they liked the week in a fancy resort with all their friendly compatriots, the reaction was, "Well, of course, who wouldn't?"

When asked what they got out of the leadership development gathering and used back on the job, the answers didn't come so quickly. Several participants commented about how much they liked the lectures, so the organizing team wondered if they needed better, higher-paid speakers. Others said they enjoyed the leadership development activities, so they went out and researched team development exercises. Yet others believed the afternoon excursions helped them connect with each other and perhaps even helped build "collaboration." By the end of the conversation, it was clear that while the event was a social success, it was less clear that it was a business success.

The new CEO saw some merit in the gathering but knew that for it to deliver its greatest value, it needed something more. He reached out to two consultants, who were experts in leadership development, to see if they were willing to fulfill his quest of finding and achieving the highest possible value.

"Could you take a look at what we are doing and retool it to create greater success?" "Maybe," answered the consultants. "But first, *we* have a few questions."

They began by asking a simple one. "What is the business purpose for having this event?" And then another: "How will you know this event has been a success and have the impact that will grow the business?"

The excitement about the event deflated when it became clear that no business case had ever been established for the annual affair. The focus had been on creating a fun-filled time that would bring people together and provide some lectures and learnings which would pump people up and hopefully carry forward into the future. There was no methodology to ensure that the leaders would apply the content or that there would be some kind of structured follow-up that would tie the leadership development learning to actual business skills which would, in turn, drive performance. Further, organizers assumed that every leader would benefit from whatever was offered without ever considering his or her individual needs. In other words, the results of this analysis led to the same conclusion as the research that most leadership development programs, while fun, do not impact business outcomes.

Not being ones to settle for mediocre business results, the consultants proposed a completely different approach. They proffered the idea of building an approach based upon sound research and evidence-based practices around leadership development that would be transformative. The one-and-done annual fete would kick off a multi-step process built upon the strategy of the business and the leadership skills critical for the strategy's success. The program would also be geared to the development needs of each individual leader, with the understanding that no two leaders develop alike.

The consultants called their approach "Reimagining Leadership Development." They promised to facilitate a high-impact process that would draw upon the new science of leadership development. This would ensure the participants would leave the event with an approach that would drive business success through the powerful use of what the consultants called "Intentional Leadership Development."

The reimagining of the annual fete was a smashing success. The organization's leaders had powerful new tools they could use for themselves and, even better, share with their teams. They developed new skills they could use right away, and they had a new appreciation of how powerful and vital their roles as leaders were for creating success for the venture. The CEO and executive team members were more than pleased with the result.

The consultants were grateful they had to opportunity to share their ideas with these leaders and help bring a new approach to how this company thinks about developing leaders. The consultants' mission was (and is) to change the way organizations approach leadership development.

Over a celebratory drink at a local pub, the consultants had a long conversation about how their approach, while impactful for the firms they work with, takes time when working with these clients one at a time. A light bulb hit them both at the same time. . . . They could remedy that with a book called *Strategy-Driven Leadership*.

B. THE STATUS OF LEADERSHIP DEVELOPMENT TODAY

Chances are that our fairy-tale scenario resonates with you—at least the beginning part, when an event without a clear purpose is planned. You could probably recount situations in which an organization invested heavily in leadership development, even when it was not exactly clear why or what impact that development was to have on the business. The development may have been based on a recent best-selling pop-psych book, a directive from an executive or an initiative by a zealous organizational development (OD) professional.

Perhaps you've been a member of the senior executive team footing the bill for an organization's annual leadership retreat and wondered if all the rah-rah speeches, team building exercises, and expensive outings were nothing more than a perk for participants. Maybe you've been a member of a planning team and felt pressure to top the previous year's event with good times, good food, and a splashy venue. Maybe you've been a talent manager eager to provide a meaningful leadership development event—but lack the access to the business information and strategy you need to accomplish this. Maybe you've been an observer when a vendor has managed to infiltrate the organization's executive ranks and sell a one-size-fits-all program that was likely to fizzle in six months or less.

If any of these roles or perspectives describes you, this book is for you. Since we first began to address the gap between leadership practices and business results many years ago, we've been successfully working with executives and talent professionals across diverse industries. We've dug deep into the research and collaborated with others who share our goal of business-driven leadership development. We are eager to share our perspective and practices with you.

*****

It's estimated that US companies spend more than $14 billion annually on leadership development (Wentworth & Loew, 2013; O'Leonard & Loew, 2012). Match that number to the abundant and growing research that finds most leadership development to be ineffective, and the conclusion is a phenomenal amount of waste.

- A 2015 study by the Brandon Hall Group found that 81% of organizations reported that they were not very effective at developing their leaders. Only 18% reported that their leaders were very effective at meeting key business goals (Loew, 2015).
- In a survey conducted by the Ashbridge Business School in the United Kingdom, only 7% of senior leaders said that their organizations effectively develop global leaders (Gitsham, 2009).
- A Corporate Leadership Council survey of 1,500 managers in 53 organizations around the world found that 76% felt that their learning and development functions were ineffective in helping them achieve business targets (Mitchel & van Ark, 2017).
- A 2017 survey by McKinsey found that only 11% of 500 global executives felt that their leadership development efforts achieved desired results (Feser, Nielson, & Rennie, 2017).
- In the 2017 CEO Challenge study conducted by the Conference Board, only 41% of global CEOs felt that their leadership development efforts were of very high or high quality (Mitchel & van Ark, 2017).
- A 2011 survey of more than 14,000 line leaders and HR professionals found that only 38% of the leaders and 26% of HR felt that the quality of leadership was very good or excellent. When asked to consider the future, only 18% of HR and 32% of the leaders saw their pipeline of talent to be strong or very strong (Boatman & Wellins, 2011).

(For an Op-Ed perspective, see "The Great Training Robbery: Why the $60 Billion Investment in Leadership Development Is Not Working" in Forbes. com, Peshawaria 2011.)

As disheartening as these research findings are, our experience convinces us that the situation is not that dire. Properly designed and executed leadership development can make a difference.

We have both been around long enough to see baby-step changes in learning and development that have been helpful. We now see an opportunity for a significant reimagining, particularly in the ability to build the leaders needed for today's unique business environment. And the timing couldn't be better. The United States and Europe, in particular, are facing a "silver tsunami" of aging leaders ready to retire . . . and a smaller supply of talent ready to replace them. It is time for a complete makeover of the approach used to create the leaders needed in today's volatile, uncertain, complex, and ambiguous business world. We call the new model Strategy-Driven Leadership Development (SDLD).

As you will see in the chapter titled "The Foundations of Strategy-Driven Leadership Development," this new approach did not pop into our minds one night over a glass of wine. It came from a wide range of insights garnered from our combined 70 years of business and professional experience. Now is a good time to give you a bit more information about that background. Then back to Strategy-Driven Leadership Development.

Richard's path started with him working on Wall Street, where he saw how the impact of poor leadership, underutilization of employee capabilities, and management overconfidence led to destructive workplaces. His observations helped him decide to pursue a PhD in psychology and to where he eventually built a successful behavioral health care business that was acquired by a national company. In subsequent years he helped that organization develop new product lines that led to a doubling of valuation and the eventual acquisition of that company by Humana Health Care.

Following the success of that acquisition, Richard was actively recruited to Pittsburgh to help grow a large health insurance business. During his time as a senior executive, Richard oversaw multiple business lines and was part of the leadership team that grew revenues from $800 million to $1.5 billion. Based on his business and corporate experience, Richard launched his own organization effectiveness and coaching consulting practice, which he has run since 2009. His unique perspective on leadership, strategy, and resilience is based on the idea that by using strengths, assets, and skills, and by aligning them in a purposeful and powerful manner, real change occurs.

Michael began with a graduate degree in industrial and organizational psychology and initially plied his trade in the steel industry in a variety of human resources (HR) and OD roles. His business leadership experience grew when he moved to a small manufacturer, where Michael headed up a business unit and led operations until the business was sold. His OD and operations experience included a healthy dose of statistical process control, Six Sigma, and lean manufacturing, creating a strong bias toward evidence-based practices. Since starting his own consulting practice in 2007, Michael has helped more than 50 organizations in a variety of industries think differently about how they develop and sustain the capability of their organizations and leaders.

The two of us frequently team up for the good of our clients. We spur each other on to keep growing, developing new tools, and driving results for our clients. Enough about us; now back to Strategy-Driven Leadership Development.

Strategy-Driven Leadership Development is an evidence-based, *deliberate, and systematic* effort that involves the following:

- Reviewing and clarifying the business strategy with all stakeholders involved in leadership development
- Describing the demand that business strategy places on an organization's leaders by translating the strategy into a few *mission-critical leadership competencies*
- Identifying and assessing the capacity of leaders at all levels to meet the demands through *robust talent assessments*
- Accelerating the development of mission-critical competencies through *Intentional Leadership Development*

No talent initiative can be effective unless it is built from a specific strategic foundation. And no two organizations have exactly the same strategy. As a result, the demands that strategy places on an organization in terms of needed capabilities or leadership competencies are not the same. Nor do different organizations start with the same existing set of talent; each organization's current capability is different.

Figure 1.1 shows how the outcome of Strategy-Driven Leadership Development is a focused process that drives business results. The expected impact is defined up front. It begins with the end in mind and builds the

FIGURE 1.1
Strategy-driven leadership development.

foundation to answer the questions, "Why are we doing this?" and "Is it worth it?"

The most reimagining in the SDLD model is reflected in our approach to Intentional Leadership Development. That's why you will find a significant chunk of this book will be committed to describing the theoretical foundation, supporting research evidence, and practical application of Intentional Leadership Development. In short, Intentional Leadership Development can be described as:

- Establishing a strong personal and business case for development
- Identifying and targeting the critical few leadership competencies which will have the greatest payoff for the individual learner and the organization
- Creating an Intentional Development Plan that is built into daily work and involves regular feedback, reflection and progress tracking, and
- Assuring that new competencies are effectively applied in real-life situations so that the developmental experiences are positive (additive) to a career, setting the stage for further development.

To simplify matters, we call these four components of Intentional Leadership Development Frame It, See It, Own It, and Connect It, or the *It Formula*.

Aligning talent strategy to business strategy as highlighted in the SDLD model has been shown to have clear positive benefits. A nationwide study of over 1,000

ON LEADERSHIP DEVELOPMENT

"Buying off-the-shelf packages and installing them without regard to VBC's (vision, brand, culture) and go-to-market strategies is frankly a waste of time and money."

(Fitz-Enz & Mattox, 2014, p. 22)

FIGURE 1.2
The business case for leadership development.

publicly traded firms *found* a positive impact of strategic talent practices on important employee outcomes and also on corporate financial performance. The study determined that a one-standard-deviation increase in high performance workplace practices was associated with a 7.05% decrease in turnover and $27,044 more in sales per employee. In addition, that small change in key practices was linked to $18,641 more in market value and $3,814 more in profits per employee (Huselid, 1995) (currencies reported in USD.)

Likewise, deliberately and systematically developing leaders has a payoff. For us, the biggest payoff (and the strongest business case) for leadership development is the effect that leaders' behaviors have on organizational culture. Research confirms that organizational culture is a driver of organization performance (Denison & Hallagan, 2017; Boyce, Nieminen, Gillespie, Ryan, & Denison, 2015), including revenue growth, market share, returns, quality, productivity, customer satisfaction, and innovation. The eminent culture scholar Edgar Schein (1996) described culture as "one of the most powerful and stable forces operating in organizations." Culture comes first; performance follows. But what drives culture? The single biggest factor is the behavior of an organization's leaders (Sarros, Gray, & Densten, 2002). See Figure 1.2 for a graphic representation.

With the advent of talent analytics, it has become easier to analyze the link between the skills of an organization's leaders and key business outcomes. For example:

- Research conducted by the Centre for Economic Performance and the McKinsey Company found that, even accounting for economic factors, well-managed companies significantly outperformed poorly managed firms in terms of productivity, return on capital employed (profitability), and sales growth (Bloom, Dorgan, Dowdy, Van Reneen, & Rippin, 2005)

- Studies that analyzed the impact of top leader self-awareness consistently found a strong positive correlation between high self-awareness and key business measures such as employee satisfaction, net profit, return on investment and stock performance (Okpara & Edwin, 2015; Wexley, Greenawalt, Alexander, & Couch, 1980; Zes & Landis, 2013)
- A longitudinal study by Russel (2001) found it was possible to predict the short- and long-term performance of executives from competency-based assessments. It also estimated that using a selection procedure based on targeted competencies generated an additional $3 million in annual profit per executive selected
- A study of 60 Fortune 1,000 firms showed that ability to lead effectively explained almost twice as much of variance in the firms' profits as did economic factors (Hansen & Wernerfelt, 1989)
- An I4CP study on talent risk management found that effectively utilizing learning and development practices positively impacted a company's financial performance (Martin & Armitage, 2016)
- A study of 72 US Army light infantry platoons found that certain aspects of platoon leaders' and sergeants' behavior were predictive of platoon performance (Bass, Avolio, Jung, & Berson, 2003)
- Organizations whose leadership development is at a strategic level (exemplary senior management support, senior leaders view leadership development as an integral part of the overall talent management system, development content aligns with strategic priorities, use of a broad learning format) exhibited 20 times greater employee retention, 12 times the business growth, and 8 times better bench strength, among other impacts, than organizations with inconsistent leadership development (O'Leonard & Loew, 2012)
- A 2013 study involving 3,000 leaders found that organizations with the strongest leadership bench exhibited twice the revenue and profit growth than organizations with a weak bench (CEB, 2013)
- As little as a "10% change in leadership bench strength leads to a 0.5% year-over-year change in revenue and profit" (CEB, 2014)

Developing the competencies of your organization's leaders can make a difference in your organization's overall performance. However, as we will describe, positive change for your organization doesn't come from any and all competencies, and it may not come from developing any and all leaders. For competency development to be effective, strategy must come first. Strategy begets strategy-critical competencies and roles which beget targeted development strategies which beget Intentional Leadership Development.

C. HOW TO USE THIS BOOK

The audience for our work on strategic talent management has consisted of three different types of leaders:

- HR/talent leaders who see the role of their function as strategic— much more than transactional or administrative
- Leaders in non-talent management roles who are interested in the practical steps in effectively developing and coaching members of their own teams
- Enlightened leaders in the C-Suite and other functions who understand the important link between talent, results, and strategy

Each of these audiences wants to have a different conversation. Executives typically want to understand the business case for our approach, along with a high-level application of the basic talent concepts. Leaders with talent responsibilities usually want to dig deeper into the foundational research, evidence about results or impact, and the how-to's.

Writing a book for these difference audiences has created some unique challenges. We wondered: Should we write two separate books? Should we ignore one set of readers and focus on the other? If so, which one? We resolved the dilemma by including something for each audience in a single tome. Our goal was not to dilute our focus by trying to appeal to everyone, but to create an accessible resource for three audiences with the same goal of developing talent for organizational success.

No matter your role, if you have limited time and want a good understanding of how Strategy-Driven Leadership Development changes the game and solves the problems associated with traditional leadership development, read Chapter 5. This chapter covers the foundation of our work on developing leaders and the 10 essentials of Intentional Leadership Development.

If you begin with Chapter 5, we hope it will pique your interest and cause you to spend more time with other chapters in the book. Here's an overview of the chapters:

- Chapter 2 touches on the status of leadership development today and the evolution of our Strategy-Driven Leadership Development model. Talent management leaders will be interested in the history, research, and theory that undergirds Intentional Development. Our approach is by no means a quick fad or superficial perspective. It's rooted in research and evidence. We are focused on reimagining for better results, not throwing out progress that has been carefully crafted over the years.
- Chapter 3 provides details on how an organization can identify the talent demands that its strategy places on their organization. It answers the question: To successfully meet our business strategy, what leadership competencies do we need?
- Chapter 4 deals with the next logical step in Intentional Leadership Development. Once the organization knows what leadership skills and competencies are needed, that organization needs to assess the leadership talent currently in place. This chapter provides guidance, based on the latest research, regarding the best tools to assess talent and how to use talent data to create high-impact talent strategies, including leadership development.
- Chapter 5 presents Intentional Leadership Development as the logical outcome of tying talent development to business strategy. The results of comparing the organization's talent demands (based on a specific business strategy) to the talent skills and competencies currently in the organization become the driver for leadership development efforts. This chapter covers the most effective ways to build those development efforts.
- Chapter 6 provides an overview of the four steps involved in our Intentional Leadership Development model: Frame It, See It, Own It, and Connect It, or the *It Formula*. These steps are essential in ensuring learning and competencies stick. The steps represent the antithesis of the one-shot development programs described in our opening fairy tale. The Intentional Leadership Development model will be of interest to talent management professionals and leaders who are looking for a solid foundation for leadership development. This chapter will be of special interest to all leaders who seek to help their direct reports grow.

- Chapter 7 explores the role of the coach in Intentional Leadership Development and provides guidance in how to be an intentional coach. This chapter explores the characteristics needed for an employee to be receptive to coaching. Once again, this chapter will be of special interest to all leaders who seek to help their reports grow.
- Chapter 8 delves into the analytics that can be used to mine all of the data that is gathered from robust and systematic talents reviews. In addition, it describes the most effective method to evaluate the impact of any training or development effort including Intentional Leadership Development.
- Chapter 9 describes the steps in becoming more strategy-driven in your talent practices as a journey. This journey to a fully integrated, supported, and effective approach to Strategy-Driven Leadership Development can sometimes become sidetracked or even stalled. The chapter highlights some typical SDLD derailers, the impact they can have on the organization, and some remedies on how these derailers can be avoided or addressed.
- Chapter 10 acknowledges that even though we began the book with a fairy tale, there is no time for fairy tales in leadership anymore. The stakes have gotten too high. This chapter reiterates the importance of Intentional Development and how leaders steeped in the skills of coaching and development will determine future success for organizations. The book concludes with a discussion of the future of leadership and how it will grow into the fabric of organizations so that development is seen as an important benefit to the organization, current and future leaders, and shareholders.

2

The Foundations of Strategy-Driven Leadership Development

Research on innovation has shown that new ideas seldom come out of nowhere as a single flash of insight. Most innovations come from the discovery of unique connections among existing ideas that are forged to create new insights. The evolution of Strategy-Driven Leadership Development took a similar path for us. It sprouted from a number of "ah-ha" connections from a variety of sources.

A general theme across our search for insights (and in this book) is the emphasis on evidence-based practices, what is now popularly called evidence-based management (EBM). Much of this emphasis came from the rigors of graduate school, from our experiences in Lean Six Sigma, and from focusing on healthcare outcomes. EBM is "about making decisions through the conscientious, explicit and judicious use of the best available evidence from multiple sources . . . to increase the likelihood of a favorable outcome" (Barends, Rousseau, & Briner, 2014, p. 4). EBM is relatively new to the practice of talent management, but we have leveraged scientific literature, the data gathered from our client organizations, our professional experience and stakeholder feedback and analysis wherever possible. Early in the new millennium, Richard and Michael both began consulting practices and began to collaborate to serve our clients. We regularly took on clients that were (1) not seeing a return on investment in leadership development at a time when the economic climate was challenging every penny spent or (2) realized that they needed to build their talent pipeline if they were to be successful. In 2008, a friend and colleague, Tim Mooney,

> If US companies spend about $14 billion on leadership development annually and 85% of it has no impact, then we are wasting almost $12 billion on development that doesn't make a difference!

published a book with Rob Brinkerhoff entitled *Courageous Training: Bold Actions for Business Results.*

Courageous Training offered several new connections and insights for us. For example, Mooney and Brinkerhoff were finding the same low impact of training that we were seeing with clients, reporting that only 15% to 20% of participants in corporate training programs ever applied what they learned in a fashion that created worthwhile results for the business.

We adopted Mooney and Brinkerhoff's recommendation to build an "Impact Map" as the first step in any developmental intervention. An Impact Map is similar to the Covey concept of beginning-with-the-end-in-mind. You start by identifying *the business and unit goals* the development intervention is intended to affect. In other words, you describe the business case for the development effort. From there, you cascade the map down through the *results* the learner must achieve to attain those business goals. Then, you set the course to the critical on-the-job actions and competencies the intervention must create.

An Impact Map helps clearly identify the business imperative and then aligns the intervention to enable the employees to achieve that outcome. Mooney and Brinkerhoff refer to this as creating "a clear 'Line of Sight' . . . a step-by-step linkage that connects training and development, job behaviors, and specific results." Brilliant!

Figure 2.1 shows how we adapted the Impact Map to focus on individual leaders. We have learners develop their own unique Impact Map as a foundation to framing their Intentional Development

Key Competencies to Develop or Enhance	Critical Leadership Challenges or Results	Key Team Results	Business Goal or Imperative
Specific competencies the leader must try out, use more of, use differently or use better to improve his/her performance	Key situations where different performance from the leader would lead to better team results	What the leader's team needs to achieve to drive the business goals	Critical few goals or imperatives needed to achieve the organization's strategy

How does this happen?   **Why is this important?**

FIGURE 2.1
Impact map for intentional development.

Source: Adapted from *Courageous Training.* Used by Permission.

Impact Maps are built from the right (Why is this important?) to the left (How does this happen?) so that the business logic for development is established in a simple, linear format. In application, learners read the Impact Map from left to right. Used in this way, an Impact Map "is an excellent communications tool to support a dialogue among stakeholders to help them understand, revise if necessary, and agree upon the logic for training" (Mooney & Brinkerhoff, 2008, p. 44).

The logic behind and benefit from developing an Impact Map is somewhat self-evident. However, we don't have to take it at just face value. Some of the best learning analytics research that looked at what predicts individual learning and eventual job and business impacts (in 326,000 training events!) found that training and development must be perceived as relevant, practical, and applicable if it is to have a meaningful impact (Bontis, Hardy, & Mattox, 2011). Development just for the sake of development does not make a difference. Learners must see a clear link between the effort and the outcome for development to have an eventual impact—a link succinctly described in an Impact Map.

Mooney and Brinkerhoff's work helped us further formulate several key principles to making leadership development *intentional*. It also aligned closely to our overall approach to SDLD. The key principles or guidelines are as follows.

- Never use training as a stand-alone, one-size-fits-all tool. This is a dead end, a waste of time.
- Do nothing until the business context is clear. Never implement training or development without confirming the business case. Begin with the question, "Why, exactly, are we doing this?"
- Always work collaboratively with stakeholders to agree upon a key business outcome. Then design an approach that has a laser-like focus on achieving that outcome.
- Make sure that you thoroughly understand the present capability of the learners and build the development intervention based on that.
- Always get the learners' (and their leaders') buy-in and commitment to training and development.

Another significant source of our insights that formed Intentional Development sprouted from the exciting recent research on the neuroscience behind human decision making, behaviors, and interactions. In particular, the NeuroLeadership Institute and its founder, Dr. David Rock, have been at the forefront of applying neuroscience principles to leadership.

The neuroscience research confirmed the impressions we had that much of what we had previously assumed about leadership development needed to be significantly revised, at least if we wanted development to have an impact. A number of findings from the brain research influenced our reimagining of leadership development, including, among others:

- Navigating a challenging life experience actually changes the brain's physical structure and how its functions are arranged; this "neuroplasticity" can occur at any age.
- The idea that individuals have different learning styles is a myth; learning best occurs by utilizing multiple learning methods, engaging multiple senses.
- For learning to have an impact, learners need to see value in a learning situation; the content must be relevant.
- Learners must "own" their development by personalizing it in a way that is important to them.
- In order to build new a new behavior, that behavior must be easy for the learner to recall.
- Our brains crave certainty, but mild uncertainty can create attention and spark curiosity. Development requires a challenge.
- Consistency is important to making new connections; all parts of development must be consistent and logically build on each other.
- Spacing learning over time is more valuable than one-shot efforts or cramming.
- Repetition is less important to learning and recall than once thought.
- Our brains are basically designed to maximize reward and minimize threats. Considering how the human brain responds to components of development is important.
- There is a difference between asking for feedback and having feedback imposed. Unsolicited feedback can be perceived as a threat. Requesting feedback is more effective.

A significant insight related to the recent work on neuroscience came from the research over the last three decades related to the distinction between having a fixed mindset versus a growth mindset, as described by Carol Dweck (2016) Now that technology enables us to watch our brains in action, researchers have confirmed that the reactions our brains have to significant new challenges can vary. Those who respond to change as a threat tend to have a fixed mindset. An individual with a fixed mindset is prone to get bogged down in details, focus on maintaining the status quo, and experience increased anxiety when faced with significant change.

Those with a growth mindset exhibit brain activation patterns typically associated with seeing change as a challenge. An individual with a growth mindset is prone to work to make things better, be open-minded, and stay focused on broader goals. People with a growth mindset have faith in their ability to develop through learning, a realistic optimism about what it takes to develop new skills, and the ability see the purpose of their work as developing their skills and abilities

It was clear to us that helping individuals and organizations to build a growth mindset should be a foundation stone to Strategy-Driven Leadership Development.

In addition to the brain science research, our new model for leadership development has benefited from the continued development and honing of competency-based talent management. Competencies provide the tool that allows a practitioner to translate strategy into the knowledge, skills and abilities required to achieve the strategy (one of the key *demands* that strategy places on an organization in the Strategy-Driven Leadership Development model). This key insight helped us to create a talent management system that adds value because it supports and drives business results.

Here are some insights we have garnered from applying competencies in a variety of organizational settings.

- Competencies provide a common language that can be used to identify and discuss key skills.
- Competencies are expressed in terms of behaviors and help people get away from thinking of leadership as a personality driven phenomenon or solely an innate capability.
- Best-in-class organizations are more likely to have defined competencies either at all levels or for select positions.
- Competencies are bellwethers of both performance and future potential.
- Competency-based selection exhibits higher levels of validity, reliability, and utility than other methods.
- Select leadership competencies are drivers of employee engagement.
- The same dictionary of competencies can be used to describe the keys to success for almost any job within any organization globally

Yet another milestone in the SDLD journey stemmed from our experience over the years with different methods of assessing talent. Michael's initial exposure to a talent review process was in his first job in the steel industry.

The company conducted regular "Functional Reviews" in which "promotability" and "readiness for the next level" of talent in every function was discussed in facilitated group discussions. Leaders were coded by different colors and posted on a large wall chart. Most of this and other talent assessment work over the years was associated with a variety of attempts to create "backup lists" as part of a succession planning processes. Not much attention was focused on how to develop talent beyond who to send to a one-week-and-done management training program.

In the 1990s, Michael conducted some research looking at the historical effectiveness of backup lists. Surprisingly, the hit rate on backup lists (how often someone on a list actually was placed in the target role) was very low—ranging from 15% to 20%—which we found was typical in other companies as well. It turned out that the time and effort spent on that type of talent review is just not worth it. A reimagining of the typical approach to assessing the capability of an organization's talent was needed.

A major revision of our assessment paradigm came from focusing on using the assessment for other purposes than "putting people in a box" or "creating lists." We began to use assessments to match development strategies to different pools of talent—a shift that reflects a growth mindset. That's why identifying and assessing the capacity of leaders at all levels to meet strategic demands is a key part of SDLD. It's also why we've devoted a chapter in this book to assessment.

Our conception of SDLD owes a great deal to seminal works that were published around the time we started our consulting practices, and which turned the world of HR on its ear. The first of these was *Beyond HR: The New Science of Human Capital,* by John Boudreau and Peter Ramstad (2007). Boudreau and Ramstad showed us how decision science can be used to bring a strategic perspective to talent—an approach they called "talentship."

In particular, their emphasis on "pivotalness" as applied to roles and skills was a game changer for us in understanding how to determine the best return on an investment in developing leaders. Boudreau and Ramstad described that while all jobs in a business may be important, few are *pivotal*. Pivotal roles are those where a small change in capability or performance can have big impact on strategy and value.

The Differentiated Workforce: Transforming Talent into Strategic Impact (Becker, Huselid, & Beatty, 2009) helped inform SDLD from two perspectives.

First, the authors' guidance to "put strategy first, not people" outlined our case for development being strategy-driven better than any other source. Building competitive advantage "means putting strategy first and developing a workforce that executes that strategy." Becker et al. also built off of Boudreau and Ramstad's concept of "pivotalness" by more specifically defining strategic positions and designing talent processes focused specifically on executing strategy.

We would be remiss if we did not highlight the significant impact that the early and ongoing research coming out of the Center for Creative Leadership had on our perspectives on adult learning and leadership development. Michael was exposed to the hallmark "Lessons of Experience" research early in his talent management career and had the unique opportunity of attending several workshops at CCL conducted by Mike Lombardo and Bob Eichinger. Besides being extremely entertaining, the duo introduced the important role that experience plays in development and the foundation research on Learning Agility, both of which inform our work to this day.

Both of us were influenced by research uncovering the power of focusing on strengths rather than weaknesses in developing organizations, teams, and individuals. This focus on strengths became a foundational building block in our SDLD model. For Richard, this emphasis began when he was a PhD candidate at the University of Nebraska. There, he came across the work of Don Clifton, who is known as the father of the "Strengths Movement."

Clifton, a professor in psychology, began asking the question, "How would it change psychology if we looked at what people (and organizations) did well instead of their faults and mistakes?" Out of Clifton's work came the book and assessment *Now Discover Your Strengths* (Buckingham & Clifton, 2001). This is most-used leadership assessment in the world today, with over 20 million people having completed the survey.

The importance of focusing on strengths is critical to building a strategy-driven leadership approach for a few reasons. First, organizations must know what they do well in order to build on those strengths to grow. Companies that recognize their core competency, for example, will outsource or even split divisions to make certain they are focused on what they do best.

When Hewlett-Packard decided to split their business into two businesses—Hewlett-Packard Enterprises to focus on new IT infrastructure and HP Inc to drive their printing business—the decision came from recognizing that separating each business unit into its component strength would result in a better focus on achieving growth.

For individuals, the same kind of approach applies. By addressing personal and team strengths, leaders and their team members identify their best talents, and people are able to be more productive and efficient. When people are doing what they enjoy and are good at, they are more productive. People also follow leaders who are focused on recognition rather than punishment. The principle of reinforcement is central to how people improve. This principle itself becomes a strategic leadership approach.

****

The connections we built over time from these varied, and evidence-based sources eventually fell together as Strategy-Driven Leadership Development and Intentional Leadership Development. As a result of following the talent evidence, we significantly altered many of our assumptions about learning and the language we employed around developing leaders. We reimagined how leadership development is framed within strategy, how talent is assessed and how to target leadership development to provide the greatest return on the time and money invested. And we reimagined the steps we take to help leaders build capabilities that make a difference.

You now understand the foundation for Strategy-Driven Leadership Development and some of the concepts underlying it—"the why." The rest of the book will delve into the "what," "who," and "how." We will spend most of our time on Intentional Leadership Development (the how) because that's where the rubber meets the road, where push comes to shove, where the magic really happens. We will highlight the 10 essentials that form the foundation of Intentional Leadership Development and will walk through the four steps in the Intentional Leadership Development Process: Frame It, See It, Own It, and Connect It.

Before we get to Intentional Leadership Development, we'll provide more details around the "what" of Strategy-Driven Leadership Development (the demand that strategy places on your organization's leaders in terms of *strategic leadership competencies*) and the "who" (identifying and assessing the capacity of leaders at all levels to meet the demands).

At the close of the book, we will provide guidance on implementing Strategy-Driven Leadership. We'll outline several "on ramps" where can begin your own strategy-driven journey and will highlight key milestones along a Strategy-Driven Leadership Development Maturity Curve. We will also highlight the talent analytics that can be derived from our approach and the best method to evaluate the impact of any training or development implementation.

3

Organizational Demand: What Do We Need for Success?

As John Boudreau highlighted in *Retooling HR* (2010), "Human Behavior is complex, but it is not random." Understanding and applying the logic generated on leadership behavior, described as competencies, is essential to optimizing your talent to drive the success of your organization. Strategy-directed behavior is never random.

The dictionary definition of competency is "an ability to do something, especially measured against a standard." While that's not bad, a more relevant definition for our purposes would be "a set of measurable behaviors related to success at work." Competencies are sets of observable behaviors, not personality traits or motives. They are the knowledge, skills, and abilities expressed in work behaviors that people choose to use. *Success at work* is an essential component of our strategy-driven model.

Why translate strategy into competencies and not something else, like personality traits? The key driver for us is (surprise) research evidence. Research reported using a very large sample of multi-rater assessments (n = 22,014) from around the globe found that competencies accounted for **45% to 64% of the variance in leadership performance** (how much differences in job performance can be explained by competencies) across

FIGURE 3.1
The strategic driven model, with focus on organizational demand.

four organization levels (Korn Ferry, 2014). The strength of this finding is important because it means talent management processes built upon competencies can exhibit high levels of reliability and validity.

Across time and industries, leaders need similar competencies to be successful. However, at any one point in a business's strategic life cycle, some skills and abilities are more important than others. It is common to see changes in key leaders as a company transitions from one strategic phase to another. This means that organization demand is a *dynamic* rather than static element. In other words, organization demand changes throughout a company's history.

For example, the skills required to be a successful startup executive are very different from the capabilities required to be a growth or scaling-up CEO. The change in the skill demands cascade down through the organization.

During Michael's years in the steel industry, almost all of the top executives came from operations (even more specifically, the "hot end" of operations, the melt shop and hot mill.) This made sense since the strategic drivers were productivity and cost reduction. Later, competitive demands in the industry drove the need to reorganize from functional structures to market-focused strategic business units. The strategic leadership skills were no longer driven by productivity and costs but by collaborating across facilities and delivering different value propositions to different market segments—much different skill sets.

When Richard's behavioral healthcare clinical and consulting business was acquired by a national managed healthcare company, they hoped to grow a national line of behavioral health care providers. Due to changes in healthcare funding mechanisms, their growth plan began to falter, and the company decided to morph into a health and wellness organization that was eventually acquired by Humana. The planned organizational demands of growing a national chain of clinical offices that had led to one set of leadership decisions had to be rapidly revised to create leadership skills requiring developing new product offerings, sales approaches, and program implementation. Evidence of that sale to Humana demonstrated that the firm had been successful in meeting the new organizational demand.

Research has found that regularly reviewing and updating the knowledge, skills and abilities (competencies) demanded by strategy has a positive benefit for organizations. Loew and Garr (2011) looked at high performing companies (in terms of retention, engagement, bench strength,

and improved business results) and found that 62% of the high performing companies reviewed and updated their competency models every two
to three years.

A NOTE ON STRATEGY

This is not a book about organizational strategy but, clearly, strategy plays
a central role in our model. We won't spend time on how to create a strategy, but it is important to describe what should be understood about strategy to assure that the rest of the process is strategy-driven.

We often find that this is a challenge for many talent management
professionals—they just are not steeped well enough in their organization's
business, markets, and competitive position to be able to build this crucial
initial link—strategy to organization demand. If they are not, and if the organization's strategy has not already clearly highlighted the role that talent will
play in the strategy, then we design and facilitate a process that helps answer
some key questions. Walking through this process with those involved in
leadership development always has a payoff. It helps highlight why this ques-
tioning/analysis should be part of the regular strategic planning process.

Here are some examples of the questions that you can ask of your organization's strategy. You can then use the output to interpret the demand
that the strategy places on the organization's present and future leaders.

- What are our key markets? Is our plan to grow, hold share or harvest
  in each?
- Who are our major competitors and what is our competitive advantage (or disadvantage) compared to them?
- What are our key performance indicators and what has been their trend?
- What are the existing and future organization capabilities (structures and processes) that the organization must possess to successfully execute its strategy? Which are declining in relevance?
- What are our core value streams and which functions play a central
  role in them?
- Does the strategy specifically address buying, building, or redeploy-
  ing talent to achieve strategic goals and objectives?
- And finally, what are the key challenges that leaders must address to
  successfully execute the strategy?

Luckily, we are finding that it is increasingly more common that talent is specifically addressed in strategic plans, to the point where specific strategic initiatives, goals, objectives, and measures are called out. After reviewing all of that, just answering the final question above (strategic leadership challenges) is all that is needed. We'll describe the role that the answers to that question (and related questions) play next.

A. STRATEGY-DRIVEN COMPETENCIES

Translating strategy into competencies need not be a complex process. It begins by identifying a source of behavioral science knowledge within the organization or involving a third-party partner who can facilitate the process. You need this expertise to leverage the competency-based research and tools to make the process effective and efficient.

The next step involves tapping into those leaders within the business who have a clear understanding about the organization's strategy, along with the assumptions behind the strategy. The design of the facilitation with these experts focuses on answering the following basic questions related to business challenges and their impacts:

1. Based on our strategy, what challenges will our leaders face at all levels over the next <blank> years?
2. What will be the impact on the organization if these challenges are successfully addressed?
3. What will be the impact if these challenges are *not* addressed effectively?

Note the emphasis on business context/the business case in these questions.

An effective facilitation design also includes some analysis, grouping, and prioritizing of the output to show a hierarchy to the results. Having identified the challenges and impacts, the organization is poised to identify and validate the leadership competencies which will help drive the business strategy.

Some of our clients have wanted to leave the establishing of strategic competencies up to HR. We always encourage them to stretch beyond that. While HR serves an important role as the facilitators of the process and keepers of the data, involving key leaders in the process helps to

- better cement the link of the leadership competencies to business strategy
- build commitment and support for the output
- create a solid "guiding coalition" of leaders as the foundation to managing the change that goes along with implementing the results

B. IDENTIFYING AND VALIDATING STRATEGY-DRIVEN LEADERSHIP COMPETENCIES

Once the strategic challenges and their impacts are confirmed, here is the next question to answer: Based on the strategic challenges, what leadership competencies are essential to achieving the desired business impact?

A variety of tools are available to match leadership knowledge, skills, and abilities to strategic challenges. Further, proprietary assessment tools allow facilitators to gather input from a sample of leaders on strategic future capabilities and then translate those capabilities to a prioritized set of competencies.

Even if you employ tools like these, we encourage you to have a facilitated, face-to-face validation discussion of the output. Don't just accept and apply the results at face value. Face-to-face validation sessions yield a greater level of detail, capture the organization's unique language, and generate stronger buy-in to the results. Building competency-based talent management systems can be a significant change for your organization. Iteration and involvement are effective change management tools. In our work with clients, we refer to this as the "validation process" and view it as an essential aspect of tying development to strategy.

For example, we worked with large regional human services non-profit whose world was changing dramatically. Demand for their services was increasing while local, state, and federal reimbursements were declining. Regulatory requirements were going through the roof, and for-profit competitors were getting into their space. Clearly, the processes and practices that had built the successful non-profit were not going to be sufficient for them to survive.

A retiring CEO and his internal successor embarked the organization on creating a new strategy and business model. What had been a set of services in silos delivered by social work experts in a stable market now had to shift to a flexible, integrated, and nimble organization led by business-minded executives. For this major strategic shift to be successful, leaders at all levels had to understand the new requirements.

New skills could not simply be imposed. We designed and facilitated robust competency modeling and validation sessions with high potential leaders so they could confirm the strategy-competency link and build buy-in to the change. It was critical to the success of the initiative that line leaders were involved in this process. It is not sufficient to have senior leaders or HR specialists dictate how the leadership development process will play out. By engaging this leadership group, we not only obtained their buy-in for the process but also helped them develop the mission-critical skill of strategic thinking.

Identifying mission-critical skills has been described as both an art and a science. We are convinced that it is more on the science side than some would have you believe. Given that the practice of using competencies in talent processes has been around since War World II, we know a lot about what works and what doesn't. As you build your list of strategy-critical competencies, keep the following recommendations in mind, which we review in more detail below.

- Don't start from scratch
- Model top-notch performance
- Model the future, not the past
- Don't simply copy others' work
- Model roles and processes, not jobs and departments
- Align and integrate all of your talent processes
- Validate the competencies
- Make the competencies come alive for every employee
- Set the stage for the competencies to be tied to development
- Keep it simple
- Keep the models up-to-date
- Build your leadership brand

Don't Start From Scratch

Many executives tend to want to develop competency models in-house, seeking to put their own spin or bias into those models. Lacking a background in behavioral science, these executives and their teams often end up creating a mashup of personality traits, values, and behaviors to try to define their success as leaders in their organizations.

Because humans are attuned to observing, understanding, and reacting to behaviors, it is important to stay at that level. In talent development, we're not trying to determine *why* humans at work behave the way they do. We are building a description of the knowledge, skills, and abilities needed for strategic sense, described in easily understood and observable behaviors.

We were introduced to a large global organization whose CEO had pushed his executive team to start from scratch and build their own competency model for the top three levels of leaders (against the advice of their chief human resource officer). It had taken them many months and involved a lot of conflict that created some bad blood among the players. After all was said and done, the CHRO asked us to take a look at the output.

When we compared this organization's homegrown competencies against validated research and global norms, we found significant weaknesses and omissions. Their "competencies" were really just a long, tedious laundry list of mostly non sequiturs (i.e., inspires excellence, demonstrates strong management skills, and drives innovation). There was no clear link between the competencies and the organization's strategy. Since they were not expressed as specific behaviors, nobody really understood what the skills meant, and they were difficult to translate into effective talent management tools. It turned out that all the time invested by the high-paid executives was wasted, and the organization's talent practices suffered as a result.

A variety of thoroughly researched and normed competency dictionaries are available. In fact, the behaviors related to success at work are the same, although some dictionaries give them different names. We strongly recommend that you translate strategy into leadership skills using an existing set of defined behaviors rather than trying to build your own. It makes the translation more effective and easier to complete. The results are well worth the investment.

Model Top-Performance

Michael worked on a project with a company that wanted to identify the skills required to be a successful store manager at their nationwide retail outlets. He completed the job analysis but was not able to see a clear pattern in the behavioral data—until the results were split by market-adjusted store performance. It turned out that high-performing store managers defined success very differently from lower performing leaders.

We've seen similar effects repeatedly in talent assessments. Lower performing/lower potential leaders are not good at describing or identifying success. As an aside, we also find that lower performing/lower potential leaders are not good at behavior-based interviewing. They may ask the right questions but are not good at combining the information into a selection decision.

Use your best and brightest as subject matter experts to identify the competencies critical for success in your organization. Otherwise, the skills of your leaders will not help drive your strategy. Don't have agreement about who are your best and brightest? You can work backward from the results of a robust talent assessment. More on that in the next chapter.

Model the Future, Not the Past

Focusing on the future is an important corollary to modeling top performance. Competencies should describe the knowledge, skills, and abilities needed to drive future success for an organization, not what success has looked like in the past. It is essential to build direct links between the organization's strategy and the behaviors critical to driving that strategy. This is why identifying the demands that the strategy places on leaders is a critical first step in the process.

For example, reconsider the large global organization we mentioned above. Their strategic success required an organization that was nothing like it had been in the past. Their future leaders needed to be able to build and navigate a flexible, integrated, and nimble organization.

Another reason to model the future is a practical one. All leadership competencies can be developed. Some are just more difficult to learn and may take more time and a wider range of experiences to improve. Defining the competencies for the future gives you breathing room to plan out leadership development and to be better prepared for succession.

Don't Simply Copy Others' Work

A foundational principle of SDLD is that different business strategies require different organization capabilities to be successful. This means that you can't just copy what another organization developed and shoehorn it into yours. It won't work because the business context is not the same. Copying a competency model from another organization will not create the needed understanding and commitment that comes via the validation process. In addition, this "copy and paste" approach skips the critical step of validating the competencies for application in your organization, which is a risky legal misstep. The research available on "what works"—what skills may be matched to different organization capabilities is valuable— but it's unwise to use this information as *the* answer. We use it as a piece of the puzzle; additional information that informs the process.

Model Roles and Processes, Not Jobs and Departments

Not every position, career level, and function should have a unique competency description. While functional and technical knowledge requirements will vary across departments, the strategic leadership competencies will not (KornFerry, 2014). Keep your focus on building links between functions and levels. Otherwise, you will reinforce existing silos, and the cross-functional movement required for successful leadership development will be hindered. During this process, you will, no doubt, get pushback from some of your team members who find themselves being territorial about their "owning" a particular job or department. We encourage our clients to communicate that the effort to model roles and processes does not mean that a job is shifting from one place to another but is merely identifying the elements of that responsibility.

Align and Integrate All of Your Talent Processes

Once you know what success looks like, systematically focus every talent tool you have on increasing the organization's capability. Leadership development is one piece of the bigger whole.

Many organizations start with building competency-based selection and promotion, since that has the biggest immediate impact and long-term payoff. Next in line is leadership development that involves providing feedback, coaching, and targeted development plans for high potentials in mission-critical roles. Leadership development begets succession management. And so it goes. Using competencies to align and integrate all your talent processes has a compounding effect and creates a common language around talent. Figure 3.2 shows how business strategy and competency models should be integrated to drive effective talent management processes.

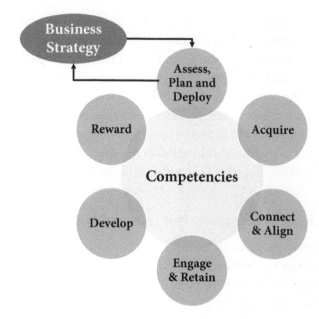

FIGURE 3.2
Integrated and aligned talent process.

Another insightful way to think about the value of integrated and aligned talent process was described by Dave Ulrich and Richard Lake in their influential 1990 book, *Organizational Capability: Competing from the Inside Out*. They describe the important role that competencies play in creating a shared mindset within an organization and that talent processes *generate* (Selection, Development), *reinforce* (Performance Management, Rewards) and *sustain* (Organization Design, Communication) strategy-critical competencies.

Validate the Competencies

As we mentioned above, we start competency modeling sessions by facilitating a discussion around three questions:

1. Based on our strategy, what challenges will our leaders face at all levels over the next <blank> years?
2. What will be the impact on the organization if these challenges are successfully addressed?
3. What will be the impact if they are not addressed effectively?

The challenges identified in answering question #1 are analyzed and prioritized before the specific impacts are determined. Then, we present a bank of competencies to the best and brightest subject matter experts. These experts rank the competencies as being essential, useful but not essential, or not important in addressing the key leadership challenges.

The results are posted for all to view, and we work to reach consensus among the experts. In the process, we refer to the competency research and norms. If we detect some gaps, we add those to the discussion. Once the competencies are assessed and selected, we build a connection from each competency back to the challenges and impacts. When a connection is unclear, we challenge if a given competency is essential. The process develops a form of content validation—confirming that the competencies are actually required in the leadership roles to meet strategic objectives.

Make the Competencies Come Alive for Every Employee

Whatever process you use to validate your competency models, make sure that those models do not get locked away in HR. Managers and employees need to understand the models and have the opportunity to interact with them. You can create opportunities by with the following initiatives:

- Build development tools
- Train managers in providing coaching and feedback
- Build self-service assessment and career planning tools
- Train everyone involved in the hiring process in behavior-based interviewing using the competencies

One of the best approaches we've seen to making the competencies come alive was an organization that used an internal collaborative site to crowd-source information about mission-critical skills. The conversations were facilitated by executives skilled in a particular competency. They asked questions and started conversations about the competency: what it was, how it was developed, how you could know if you were skilled or not, and examples of people using the competency effectively or ineffectively. The discussions became active and robust; the competency came alive for this global organization.

"The point is to match your development resources with assessed deficiencies. One-type-fits-all leadership models are a terrible misuse of time and resources."
(Fitz-Ens & Mattox, 2014, p. 24)

Set the stage for the competencies to be tied to development: Competency development is challenging work, and leaders fairly ask if the investment of time and effort is worth the payoff. Tying the act of the competency modeling to individual development helps your leaders begin to see where they can apply the competencies to their own efforts. The value of competency modeling, and eventually development, is not just about the success of the organization. It is also in the efficiency of the work that gets completed by the leader. A common example is around the competency of "delegation." We often work with senior leaders who have difficulty letting go of key responsibilities. As a result, they are put in long hours and find their efforts being less effective than they like. Discovering how to be a good delegator helps these leaders run their businesses better and gives them some time to go to their daughter's soccer game.

Keep It Simple

Skills are most likely to be retained and used when initiatives focus on a few mission-critical competencies that fall under themes at a time. The parts need to fit together in a foundation that is easy to understand, recall, and apply. These ideas, which come out of recent research in neuropsychology, suggest that the human brain processes information more effectively when easy-to-remember terms (or mantras) are used to help people understand and remember these concepts.

With many of our clients, we share the "rule of three," which states that people can easily remember three key elements of an idea or issue. Framing strategy, sales, or decision-making criteria with three key elements creates a high probability that learners will integrate the elements into a whole.

For example, at the NeuroLeadership Institute's 2017 Summit, a presenter from Hewlett Packard (HP) highlighted the following three simple, easy-to-remember mantras designed to clarify expectations for all leaders as the company split into two new companies.

Imagine the future
Inspire the team
Make it happen

Similarly, at the 2019 HRP+S Annual Conference, the VP and chief people officer of Southwest Airlines outlined the three components of "Living the Southwest Way."

Warrior Spirit
Servant's Heart
Fun-LUVing Attitude.

Both presenters went on to describe how each firm used these simple structures to organize a basic set of leadership competencies supporting the strategic themes.

Keep the Models Up-to-Date

As the competitive pressures change for your organization, so will the requirements for success. As strategies change, the demand on the organization and its leaders will change. Review your competency models anytime you update strategy. The impact of applying the competencies should also be tracked to see if any improvement is required.

We've been working with one client for more than six years now, and while our role with them evolved from doing training to advising, we've returned to review and modify their competency model three times over that six-year period. As their team developed new skills and strengthened their business acumen, the competency requirements shifted. At first, we were focused on basic operational activities such as planning, aligning, and communicating. Now our work has focused on developing business acumen, building effective teams, and creating strategic agility.

Build Your Leadership Brand

We have a consumer retail client that has a very distinct and well-articulated brand in the markets where they compete. When it comes to rolling out talent management initiatives, their consumer insights and marketing group is called in by talent management to help brand the processes internally. Their internal branding matches their external branding and makes a real difference in buy-in, commitment, and understanding of the new processes and systems. We would encourage you to think about validated competencies as your "leadership brand" and to market it internally as such. The competencies reflect how your leaders help your company compete and what makes your culture unique. Your leadership brand will help drive engagement and retention.

SEEING THE FIELD AT THE RIGHT TIME

When General John Buford of the Army of the Potomac arrived in the small central Pennsylvania town in Gettysburg on June 30, 1863, he had already discerned that the Confederate army was fewer than 50 miles from his location and that the next great battle of the Civil War would be there. Buford and his cavalry leadership surveyed the potential field of battle and recognized that he needed to secure the eastern high ground south of Gettysburg so as to give the rest of the Union army the best chance to win the battle. In a series of small skirmishes before the main battle, Buford was able to keep back the Confederates until the rest of the Union Army arrived.

Buford's ability to "see the field" and determine the strategic importance of location is considered to be one of the key elements to the ultimate victory in that battle. Buford's great contribution was not in his fighting of the pre-battle clashes (which, of course, he engaged in with his troops) but in how he aligned his troops into what was an actual defeat into an eventual victory.

Whether Buford was born with his strategic ability or it was groomed during his time at West Point is much less important than the fact that he used his talent at the right time and most certainly in the right place.

Responding to the demand the organization needs at the time it needs it most critically is a hallmark of great leadership. By developing the strategic competencies in leaders, organizations are better prepared for their great battles as well.

4

Assessing Organization and Leadership Capability

Now that you have an idea of what the organizational demands the business strategy puts on the business, we can now turn to identifying and assessing the capacity of leaders at all levels to meet the demands placed on an organization by its strategy, a critical component of the Strategy-Driven Leadership Development Model.

The purpose of assessing your talent is twofold:

1. Understanding the overall capability of your talent and whether you have the talent necessary to execute your strategy
2. Identifying the most effective development strategy for similarly situated employees—identify the individual leaders and/or pools of talent that would benefit most from an investment in their development, thereby providing the best return for the organization

Talent assessment, like leadership development, has evolved over the years. In this chapter, we'll describe the current state of that evolution by looking at the two most common criteria used in assessing the capability of an organization's talent: growth potential and performance effectiveness

FIGURE 4.1
Strategy-driven leadership development, with focus on leadership capability.

over time. You can find many different versions of these two criteria (right stuff versus right results, what versus how, etc.). Our assessment paradigm reflects an evidence-based approach.

A common axiom states that past performance or behavior is the best predictor of future performance. ***But this is only the case when the future situation is the same as the one encountered in the past.*** What's more, performance is only part of the talent capability equation.

Many organizations over-emphasize past performance when they assess the capability of their talent. Performance is only part of the talent capability equation. For example, a 2005 study by the Corporate Leadership Council (CLC) found that 71% of leaders identified as high performers were not also assessed as being high potentials—because they didn't have the agility and capability needed to move up.

If these results hold true in your organization, moving people to roles requiring more scope and scale based solely on their past performance may reduce, rather than improve, the overall capability of your organization. Interestingly, the same CLC study found that 93% of those identified as high potentials were also seen as being high performers. The development strategy for high performers needs to be different than that used for high potentials.

With our emphasis on identifying the most effective development strategy for similarly abled employees, it makes sense for us to consider both factors of potential and performance.

WHERE AND HOW TO INVEST IN DEVELOPMENT

Consider if you had the following four people in your organization:

- Elon Musk
- Albert Einstein
- Pat Smith (one of the many solid, engaged performers who is happy in his or her current role and that make up most of your workforce)
- Milton the Stapler Guy (from the movie *Office Space* who ends up burning down the company over want of his prized stapler)

From whom would you get the best return on investment in leadership development? And what development would you provide?

We'll start with the easier answer. There's no return on investment in Milton's development beyond eliminating a barrier or risk to the company's success. The only investment would be to help Milton move on to another job with another company—safely, legally, and compassionately. For Pat, the focus for her should be on keeping her engaged and up-to-date. The best investment would be supporting her technical and functional skills development which she would self-manage.

The real payoff comes from investing time and resources on Albert and Elon, but the most effective type investment would be very different for each. You need to retain Albert and have him share his bountiful knowledge with as many people as possible internally. He should have a strong network at all levels inside and outside the company, and you should prepare for his succession through focused knowledge transfer. For Elon, you need to make sure that he is constantly challenged and intentionally learning through placement in a variety of challenging, mission-critical roles. He should be exposed to the top leaders of the company and to other high-performing, high-potential employees. He should be developed *intentionally*.

Do you know who the Elons, Alberts, Pats, and Miltons are in your organization? How are you differentiating the investment in their development? Or are you taking a one-size-fits-all approach?

A. WHAT TO LOOK FOR: GROWTH POTENTIAL

Everyone wants to know the secret of success, and there is one. It's called continuous learning to do what you don't know how to do.
—*The Leadership Machine*, 2011

In a time of drastic change, it is the learners who inherit the future. The learned usually find themselves equipped to live in a world that no longer exists.
—Eric Hoffer (1898–1993), US philosopher

A variety of criteria have been used over the years to describe an individual's potential to move into larger and broader roles. The most common criterion was the idea of "promotability," which was described as how many levels above a person's current role that individual could progress.

We dislike thinking about potential this way because promotability is influenced by the availability of vacant positions or by the number of boxes on the present-day organization chart. What about positions that don't exist now but are foreseen in the strategy? What about strategic assignments or projects that don't show on the chart because they aren't full-time positions?

The concept of promotability also does not match well with less hierarchical organizational structures such as matrixed organizations or strategic business units. In addition, promotability typically has no clear criteria to define it; it is often just the opinion of one boss. In other words, promotability has most often been based on subjective rather than objective criteria.

"People high in learning agility tend to take control over their own learning by looking for opportunities to grow, requesting feedback about their work, and continually engaging in self-reflection and evaluation about their work and careers. They learn quickly, trust themselves enough to experiment with new solutions, and apply their new knowledge to novel situations. Unsurprisingly, this means agile learners deliver results for their organizations."

(Cavanaugh & Zelin, 2015, p. 4)

We prefer to stay away from assessments that reflect the organizational chart and think about potential from another perspective. Growth potential is not about whether a higher-level job is available or if an individual is ready to move to the next level. It's about an individual's ability over time to adjust to new demands, learn quickly from new experiences, and build his or her repertoire of leadership competencies.

An organization's structure can change quickly and, often, unpredictably. By assessing growth potential, we are able to better understand how leaders at all levels are ready and able to pivot to meet those changes.

We normally use three different factors to describe the current level of an individual's growth potential:

- Learning Agility
- Capability
- Aspiration

These three factors are closely aligned and inform each other. Of the three, learning agility is the most objectively measured and has the most evidence behind it as a strong predictor of job performance following a promotion. Our assessment of growth potential recognizes the interaction of these elements. Let's describe the three in more detail.

Learning Agility

In 2000, Michael Lombardo and Bob Eichinger published a journal article in which they coined the term *learning agility* which they defined as "the willingness and ability to learn from experience, and subsequently apply that learning to perform successfully under new or first-time conditions." Learning agility "focuses on human behavior, high-level cognitive processing, and the selective transference of lessons learned in one setting and applying them to a uniquely different one. It includes experimentation, self-reflection, leveraging individual strengths, continuous improvement, mindfulness, and mentally connecting experiences obtained in one situation to different challenges in another" (DeMeuse, 2017, p. 268). A recent study found it to be the most frequently used criterion (67% of organizations of all sizes, 81% in companies with 1,000–5,000 employees) to measure leadership potential (New Talent Management Network, 2015).

People who are continuous learners are described as being learning agile. Agile leaders not only navigate a wide range of challenging experiences but exhibit a pattern of learning from them. Learning agility is unrelated to gender, age, education, ethnicity, or station in the world (Lombardo & Eichinger, 2011; Dai, De Meuse, & Tang, 2013), is a significant and positive predictor of leadership competence, career success, and higher compensation (Conner, 2011; Dai et al., 2013), and has a strong relationship to both leader performance and leader potential (DeMeuse, 2017). Dries, Vantilborgh, and Pepermans (2012) studied the relationship of performance and potential and found that high performers were three times more likely to be identified as high potentials than lower performers. However, employees who were high in learning agility were 18 times more likely to be identified as high potential.

It turns out that a successful career is more than moving up the rungs in an organization's hierarchy. A successful career involves moving around and across functions and experiences and, hopefully, learning as you go. A successful career requires letting go of competencies that are not relevant to new challenges, preserving skills that can still make a difference, and adding new competencies demanded by the new opportunities.

In other words, with learning agility we are identifying what Satya Nadella, CEO of Microsoft, called the "Learn It Alls," not the "Know It Alls" (Majdan & Wasowski, 2017).

It is possible to identify people who exhibit learning agility as it involves behaviors that can be observed. Learning agility is not like IQ that is measured by a test and thought to be fixed; it is expressed in specific leadership competencies that can be learned over time (Mueller-Hanson, White, Dorsey, & Pulakos, 2005; Silzer & Church, 2009).

THE ORGANIZATION CHART IS DEAD!
LONG LIVE NETWORKS!

I (Michael) first read about the concept of "white space on an organization chart" in the 1990 book by Geary Rummler and Alan Brache, *Improving Performance: Managing the White Spaces on the Organization Chart*. The importance of organization white space came rushing back to me recently after working with technical leaders, engineers, and program managers at two global manufacturers. We were discussing what it takes to be successful as a technical leader, now, and into the future. It dawned on me that organization charts no longer even come close to depicting the way we work (even if we add dotted lines!). Most of the work gets done in the white space—the processes and projects that occur across departments and that never show up on a chart.

Organizational Network Analysis research (particularly at Carnegie Mellon University) has further confirmed what Rummler and Brache proposed. How work gets done in an organization looks nothing like the org chart . . . and executives are not good at understanding how the networks in their organizations operate (Krackhardt & Hanson, 1993).

Who-reports-to-whom doesn't make much difference when most employees really don't just have a job in a department anymore. Their critical contributions come from the roles they play in key business processes (which could be several), and on project teams (which are usually multiple). Technical leaders don't have a single boss anymore. Employees may spend more time with process and project leaders than with their managers. Managers of major product platforms don't have anyone reporting to them but must accomplish mission-critical customer projects through multiple cross-functional teams.

Rummler and Brache wrote about effectively managing the white space on the chart. We believe the additional challenge in this millennium is helping employees navigate the white space successfully. The background has become foreground.

The success factors for leadership have changed. Some employees are okay with not having a clearly defined box on a chart. They are comfortable when things are up in the air, can shift gears comfortably, and are flexible and adaptable. They are agile learners who know how to get things done in the organization and can manage by remote control and communication. These employees can influence others in multiple ways. The problem: these competencies are rare skills that are difficult to develop. It is more common for employees to be comfortable with a box, a single boss, and stability.

So what's a company to do? Send the org chart to the dungeon! Long live "Network Descriptions of Multiple Role Charts." Ummm, that doesn't work

. . . how about "Un-Organizational Charts." That wouldn't sound good to the investors . . . how about ". . ."

The point—let's begin by communicating that work is complex and things will be ambiguous at times. The message starts by changing the competencies on which we hire, coach, and develop critical talent. Management by annual objectives may not always be the best tool. Our compensation and reward systems need to change from being job-based to recognizing role and project contributions.

The Organization Chart is Dead! Long Live <you fill in the blank>!

As described by Lombardo and Eichinger (2011), Swisher (2016) and others, high learning agile people demonstrate their agility in a number of dimensions:

- **Results Agility**: They perform effectively in first time challenges, situations where not everyone would be a high performer. They don't give up easily. They are driven and inspire confidence in others.
- **Change Agility**: They demonstrate curiosity as well as comfort with complexity, ambiguity, change, and new challenges. They are calm under pressure, tend to push boundaries, and like continuous improvement
- **Mental Agility**: They demonstrate strong critical thinking and logical, methodical problem solving. They enjoy exploring new things, making unique connections, and can explain their thinking to others.
- **People Agility**: They relate well to all kinds of people. They assess situations and people quickly, can adapt their behavior to match situations, express their opinions and are likely to challenge other people
- **Self-Awareness**: They know themselves and their impact on others. Learning agile people are open to feedback and improving themselves. They learn from their own and others' mistakes.

Learning agility should play a major role for any organization when it comes to assessing leadership potential. We would agree with the authors of a 2009 review of learning agility who state that "whenever there is an effort to identify individuals with potential, it inherently suggests that the person does not currently have the end-state skills and needs to further develop to obtain them. The learning dimensions are the gatekeepers to learning those end-state skills. Without them little development or growth will occur, for any career path" (Silzer & Church, 2009, p. 41).

Capability

This second critical factor in assessing potential can be defined as an individual's demonstrated ability to handle roles or assignments with increasing depth and/or breadth. The Lessons of Experience study conducted by the Center for Creative Leadership (McCall, Lombardo, & Morrison, 1988) and subsequent research found a pattern of challenging roles or assignments that successful leaders experienced and learned from.

Lombardo and Eichinger (2011) describe a number of these "developmental jobs," which include assignments such as moving from a staff role to a line role, fixing something that's broken, taking on a role with a heavy strategic demand, managing a crisis, starting something from scratch, and chairing a task force or project. In assessing someone's capability, we look for examples in his/her work history where s/he successfully negotiated these kinds of scope and scale challenges. For example, we recently conducted a talent assessment for a client and looked at the backgrounds of those seen as high potential successors for their executive positions. One of the top-assessed candidates had experiences outside of the company in consulting and banking, had worked internationally, had led the expansion of an international business into a new market, and held roles in a variety of functions for the current organization including operations, business development, human resources, and IT. Her current role was starting up a new function focused on innovation.

In Intentional Development, we also work with learners to analyze if their current role is developmental and/or to look for assignments that would add critical competencies to their leadership repertoire.

Aspiration

The third component in assessing potential is an individual's interest in and willingness to seek out roles that increase his or her contribution and development. Some leaders are satisfied at their current level or may have met their personal career aspirations (like our example of Pat Smith mentioned in the previous section). Others consistently seek out new and different roles that not only increase their own capability but also contribute more value to the organization. We often find leaders who exhibit learning agility and have been through a variety of experiences but, for one reason or another (typically personal), do not aspire to take on more scope or scale.

Aspirational goals need to be checked out regularly with an employee as circumstances may drive their capacity for more responsibility during certain periods. For example, new moms or dads may decide that their new family member requires a greater priority than their career, but this may be a transient decision, and assuming that their family demands will always outweigh work desires may be done in error.

Recognizing this more personal component of growth potential helps assure that you add a practical, personal component to talent assessment, which is typically appreciated by the individual. The 2015 New Talent Management Network research project cited above reported that "ambition" was the second most common factor consider in determining potential, after learning agility.

B. WHAT TO LOOK FOR: PERFORMANCE EFFECTIVENESS OVER TIME

Like many of the talent paradigm shifts we are describing in this book, the paradigm of what constitutes effective job performance is rapidly evolving. Much of this has been influenced by the research on social networks in organizations and the benefits of aligning organization goals and objectives with employee objectives.

Network analysis has contributed significantly to our understanding of how things get accomplished in an organization–in the white space on the organization chart we mentioned before. A growing body of network analysis research suggests that the quality and type of connections between individuals on a team or in an organization is directly related to performance. For example, researchers at Penn State and State University of New York found that teams with densely configured networks attained their goals better than others. In addition, teams with leaders who were central to their intergroup networks performed better (Balkundi & Harrison, 2006). A study of 52 bank branches by researchers at Carnegie Mellon University found that the profitability of a branch could be predicted by characteristics of the social networks at the locations—the greater the "advice seeking" that occurred across employees of different levels, the higher the branch's profitability (Sarkar, Fienberg, & Krackhardt, 2010). In an earlier study, some of the same CMU researchers found that turnover had more to do with the relationships among those that left a company than with other organization factors. When someone in a group's network left, "others left in droves" (Krackhardt & Hanson, 1993).

The science of network analysis is advancing rapidly, but it's not at the point where we can use it directly to understand the capability of an organization's talent to achieve its strategy. We can, however, look for examples of network contribution and look beyond basic performance reviews in a talent assessment process. We have revised the performance effectiveness criteria in our talent assessment process to target:

- How effectively employees meet expectations over time and in different roles
- How much they actively contribute to the success of others,
- How well they align their work with others on their team or in their network

CEB (2013) calls this combination of individual task performance and network contribution *enterprise contribution*. Their research shows that companies that were able to move more employees to high levels of enterprise performance realized a 10% improvement in profitability compared to 5% for companies that achieved individual performance improvements alone.

There is also ample evidence of the value of aligned employee and organizational goals and objectives. For example, in a comparison of companies with strong financial performance (in terms of return on equity, revenue growth, and net income) versus weaker performers, 44% of the stronger performers were found to have 100% aligned goals at the manager level while none of the weak performers exhibited any alignment (SAP SuccessFactors and Workforce Intelligence Institute, 2006)

Overall performance, therefore, is not just related to how well an individual employee does his/her job but how much that individual contributes to the success of others and links what that individual does with others in his/her network.

As we look to reimagine Performance Effectiveness Over Time, it is easy to see that small changes in how we approach this important assessment exercise does not require a wholesale change but does demand a new way of viewing how we typically think of workplace performance as we assess and move people into leadership roles. Another paradigm reimagined.

C. HOW TO FIND IT: THE BEST ASSESSMENT TOOL OR METHOD

We've outlined the factors to look for to best understand the present capability of your leaders, but what methodology can you and your team use to best evaluate or assess Growth Potential and Performance Effectiveness?

A variety of approaches have been used for assessing talent in organizations. To be effective, any tool must meet four basic criteria.

1. Reliability: It provides consistent results across people and time
2. Validity: It provides information that is well-founded and has a relationship to key business results
3. Face Validity: The processes and results easily understood and accepted by sometimes skeptical executives
4. Utility: The results are valuable enough to warrant the time and effort invested

The Table 4.1 provides a relative comparison of what we know about the validity, reliability, face validity and utility of the most common talent assessment tools.

Single Manager Assessments (or Performance Review Ratings)

Many biases creep into the ratings managers give their employees, particularly if those ratings factor into compensation decisions. In some cases, too much familiarity limits objectivity. In many cases, managers are not comfortable in confronting negative behaviors or recognizing positives. In

TABLE 4.1

Comparison of Assessment Methods

	Validity	Reliability	Face Validity	Utility
Single Manager Assessments	Low	Low	Okay	Low
Testing or Executive Assessments	Can Be	Can Be	Varies	Can Be
Multi-Rater Assessments	Can Be	Can Be	Can Be	Medium
Structured Interviews	Okay	Okay	High	Okay
Assessment Centers	High	High	High	Low
Talent Reviews	High	High	High	High

addition, managers may worry that a poor rating might reflect negatively on their own performance as a leader.

Our own analysis of several organizations' single-manager performance review ratings led us to conclude that that the ratings revealed more about the capability of the raters than that of the ones being rated. This conclusion was confirmed in a large sample study (n = 4,492) conducted by Scullen, Mount, and Goff (2000) that consisted of managers who received ratings from seven raters (two bosses, two peers, two direct reports, and a self-rating). The results indicated that raters are biased and unreliable and that particular tendencies of the managers accounted for well over half the variance in the performance ratings! More recently, a study by Deloitte Consulting, Bersin found that performance management is "universally despised" with a net promoter score of −60, leading the authors to state that "the extent to which respondents disdained their PM (performance management) approach was shocking" (Enderes & Deruntz, 2018, p. 2).

Whatever performance reviews may mean for individuals, there seems to be little value to only using single-manager assessments to understand the capability of your organization's talent. In fact, a study sponsored by the Institute for Corporate Productivity and the Center for Effective Organizations found that the use of traditional ratings was a negative predictor of organizational performance—the more the use of the ratings, the lower the organizational performance (Ledford & Schneider, 2018). Unfortunately, we find these unreliable, invalid ratings are the most commonly used assessment tool. The Deloitte Bersin study (Enderes & Deruntz, 2018) found that 96% of the surveyed organizations still used formal reviews. Fortunately, there is a strong and growing trend to eliminate ratings and significantly revise the focus of performance management processes. (See the sidebar "Down With Performance Reviews" in Chapter 5.)

Beyond the many failings of manager performance ratings, it seems to be logical to promote employees who are the best performers in their current roles. Unfortunately, research continues to conclude that performance in a current role alone is not a good bellwether of success at a higher level (Connor, 2011; De Meuse, Dai, & Hallenback, 2010; Lombardo & Eichinger, 2000).

Multi-Rater Assessments

We are big fans of multi-rater or 360 assessments and have used a variety of very effective ones over the years. But we recommend these assessments *only* as sources of feedback for development purposes and *only* when the feedback is at the request of the learner. Biases that affect reliability and validity tend to creep into the multi-rater results when the raters know that the results might impact someone's career or compensation. In addition, the burden of completing a large number of 360s at a given point can be a logistical and practical challenge.

Testing or Executive Assessments

Testing can be a very reliable and valid tool to assess human capabilities. However, poorly designed and implemented testing can do more harm than good.

Probably the worst culprit among assessments is the use of personality testing in talent decisions. Multi-year meta-analysis of a century's worth of data has shown personality testing to be one of the least predictive of job performance (Schmidt & Hunter, 1998). Numerous issues have been raised about personality measures including validity, relevance, discrimination against those with disabilities, and being overly psychological and not business-focused (e.g., Burke & Noumair, 2002; Cappelli, 2008). The problem is compounded by the fact that there are more than 2,500 different personality tests on the market, and many have received very little professional scrutiny (Paul, 2004).

Comprehensive executive assessment can also be both time-consuming and expensive. We recommend this type of assessment be used for targeted situations and typically only at higher levels, not to broadly understand your organization's leadership capability at all levels or to confirm the best development strategies needed by similarly abled pools of talent.

Assessment Centers

Assessment centers have been shown to be a valid, reliable, and content-valid method to assess talent (Thornton & Gibbons, 2009). It's challenging, however, to gather the resources required to build and maintain such centers internally. Of course, third parties conduct assessment centers for a fee. Since using these centers is expensive, organizations tend to be selective on who participates. The challenge then becomes selecting the criteria to identify high potentials to attend. This defeats the whole purpose of using assessments to identify high potentials. Assessment centers might be best used to confirm the results of a talent review or provide additional feedback as part of an Intentional Development plan.

D. OUR ANSWER: TALENT REVIEWS

One assessment tool has withstood the test of time—a formalized and standardized Talent Review. We haven't found a reference to confirm where talent reviews started but we were first exposed to them in the 1970s. In this section, we'll describe how we've significantly updated talent reviews and will outline the best practices for assessing talent

A multi-pronged research project (using in-depth interviews and qualitative and quantitative surveys) found that having a robust and consistent talent review process that generates high quality and reliable information is a "gold-standard" best practice in succession management (Lamoureux, Campbell, & Smith, 2009). The same study highlighted that continuously aligning talent review discussions with strategy was also a best practice and helps assure that talent reviews are not seen as just an HR exercise.

Unfortunately, talent reviews are not the most common assessment tool in use. Using a sample of public and private, for-profit and not-for-profit companies, Hagemann and Mattone (2011) found that the number one method used to identify high-potential talent was the opinion of senior leaders, followed by performance appraisals, and then a formal talent review process. Given the evidence against the effectiveness of single-leader assessments and performance appraisals, the best of these top methods is clearly the talent review.

ONE GUY'S TALENT REVIEW JOURNEY

One of Michael's first corporate assignments out of graduate school was to develop an assessment center to identify potential first-line manufacturing supervisors. As described above, the centers demonstrated strong validity but ended up to be too resource-intensive for a large, multi-site organization. Michael then validated a paper-based assessment to identify future leaders. While this method also showed strong validity, it lacked face-validity (managers didn't understand how to interpret the results) and reliability (locations tended to use the assessment inconsistently or incorrectly). At the time, the facilitated talent reviews, called "functional reviews," showed the most promise.

The functional reviews morphed over time, particularly as the company was exposed to the work on leadership potential being done at the Center for Creative Leadership and when talent management software became available. The company used software called ExecuTrack, which had an active user community where they learned more about how others were using talent reviews and how talent profiles and analytics were generated from the assessments. When the headquarters of the manufacturer moved to Pittsburgh in the late 1990s, they also benchmarked the talent review process against other local manufacturers, especially PPG's global approach. They continually updated the process, particularly the criteria for assessing performance and potential and employing the Performance/Potential Matrix (9 Box) to display assessment results. (More on the 9 Box later.)

Another addition to the evolution of our thinking on talent assessment came from working with Dr. Charlie Bishop. Charlie's approach, based on the work he did at FedEx, Baxter Healthcare, Quaker Oats, and ADT, is summarized in his 2001 book entitled *Making Change Happen One Person at a Time: Assessing Change Capacity Within Your Organization*. Dr. Bishop's work not only added to our understanding of assessing talent (he emphasized Change Capacity as determined by assessing Versatility and Change Response) but how the results could provide a snapshot of the overall capacity of an organization to drive change.

As we have described, research on neuroscience, organizational network analysis, and learning agility has informed our upgraded approach to talent reviews. The talent review process we follow helps uncover what Rob Cross and Laurence Prusak (2002) called "the people who make organizations go—or stop." This is reflected in the assessment of Performance Effectiveness, where we look not just at accomplishments over time but, more importantly, at how much the individual supports and contributes to the success of others and how well he or she aligns his or her work with the work of others and with broader business objectives. In addition, our talent review approach leverages what we now know about Learning Agility to help uncover those leaders with potential (and to enhance Intentional Development).

It turns out that the talent assessment evolution we experienced mirrored the journey that many of our clients have traveled. They also concluded that facilitated talent reviews are the best method to assess an organization's talent. One client had historically used a variety of assessment methods for a variety of reasons including executive assessments, multi-rater assessments, and talent reviews. They also did annual performance reviews. When they decided to pare down their assessment tools and compared the many approaches, they concluded that their talent review process provided the most robust overall information on their talent pipeline and had the broadest support from their leaders. (Surprise, surprise—performance reviews were seen as the least useful.)

E. HOW TALENT REVIEWS WORK

When designing and implementing a talent review process to assess an organization's leadership capability, we seldom follow the exact same path. Each organization is at a different place along a maturity curve in their strategic talent management processes, so we to start at different places with different tools.

For example, one of our clients had previously done talent reviews using paper-based assessments done by single managers. The reviews were focused on creating backup lists for succession purposes. Their strategic business case for spending time on talent reviews was pretty solid, and they also had begun using competencies for selection—but not across the organization.

Our on-ramp for that client focused on revising the approach to gathering talent assessment data, integrating competencies into the process, adding a calibration step to assure consistency, and using the results for development and planning, not just for backup. Their HR department was pretty savvy (meaning they had a strategic attitude toward how HR could contribute to business success), so it was easy to work with their talent professionals to get the revised process locked in and supported internally.

Compare that organization to a client with whom we had just helped revise their core business strategy and model. They were building their talent infrastructure and now they had a clear business case for understanding the capacity of their talent. Unfortunately, they had a more basic and tactical HR team with no real talent systems in place to drive the strategy.

For the second organization, our approach was much more basic and required bringing their HR team along by conducting the sessions and modeling the steps. We first introduced a standard competency dictionary and worked with the leaders to translate the strategic demands into mission-critical competencies. Next, we conducted structured talent reviews of second-level leaders with all the top leaders in the room so that (1) we could get multiple perspectives on performance and potential, (2) they would learn the process together, and (3) they would jointly learn how to recognize the key

FIGURE 4.2
Talent review process.

competencies. In addition, with the top leaders in the room, we could calibrate the assessments as we went, working with those leaders to create a common understanding of the criteria. We then analyzed the data, reviewed the results with the top leaders, and created detailed organization development action plans from the data. The next time around, we were able to streamline the process for the top leaders and to cascade the process down through the organization to the first line. We also worked with HR to develop the capacity internally to facilitate the process and build their analytics capability.

The best path forward may be unique to your company, but here are some best practice guidelines to help create a clear map, which we describe in more detail below.

- From strategy, all things must grow
- It's all about development
- The secret's in the discussion
- Just the facts
- Calibrate, calibrate, calibrate
- Performance and potential are different
- The more the merrier
- Don't climb the Tower of Babel
- What gets measured gets improved
- Once is not enough
- HR is not the owner
- Mine the data
- Take action, build accountability

From Strategy, All Things Must Grow

As we have previously emphasized, the picture generated from a talent review only has meaning if that review is done in the context of the organization's strategy. The choices that a company makes in setting the nature and direction of the business will have a direct effect on the type, mix, and structure of talent needed for success. Therefore, talent reviews should begin with a review of the strategy. A key part of that discussion should involve identifying the roles and functions that are critical to the strategy's success. (We'll touch on the role of roles in a later chapter.) If the talent review is handled as a one-off, stand-alone effort that is not integrated with your organization's strategy (or with other talent initiatives), then the process will have limited impact (Avedon & Scholes, 2010; Lamoureux et al., 2009).

It's All About Development

The goal of talent reviews is not to put everyone in a box or to create backup lists, which is how most organizations use talent reviews. The ultimate outcome is to identify the best development strategy for pools or feeder groups of talent up and down the pipeline and to allocate scarce development resources to the talent that will benefit the most (Charan, Drotter, & Noel, 2011; Kesler, 2002). Research has shown that integrating development planning with succession management is a hallmark of high-impact succession programs (Lamoureux et al., 2009). In some cases, one component of the development strategy is to confirm the individual's assessment, i.e., can we provide the individual with an experience that confirms what we have seen as his/her potential to grow?

The Secret's in the Discussion

Effective talent reviews require candid discussions about the key players in the organization. The discussions must be interactive assessments. They should not consist of solo presentations by each manager or a monologue prepared by HR. The best perspective on talent comes from a facilitated discussion among several managers based on what they have actually seen the employee do and achieve and the impact those behaviors had on others or on key business results

Effective talent reviews are akin to behavior-based interviewing with multiple raters. When an organization is in the early stages of their talent assessment maturity, we often suggest that they don't prepare for the discussions and, in particular, that they *don't* refer to performance review documents. As leaders and the organization mature in the process, we may provide tools for them to prepare for the talent reviews and to streamline the process. We find that leaders appreciate the fact that they don't have to prepare, which actually heightens their anticipation and creates a kind of excitement about the process.

For clients who are just beginning their strategy-driven leadership journey, the first go-round on talent reviews are sometimes met with skepticism. Reactions such as, "You mean you aren't going to interview every single person?" "Why aren't we using our performance review ratings?" or "Why aren't you referring to the DISC assessments they completed?" are not uncommon. Invariably, after the leaders experience the process, the skepticism quickly fades away. They appreciate the efficiency and objectivity of the approach.

Just the Facts

Effective talent discussions focus on employee behavior and its impact and avoid descending into psychoanalysis, trying to figure out why someone is like s/he is. The facilitator plays a critical role in keeping the discussions focused on behaviors that are critical for success at all levels in the organization. The focus is on what has been observed about a particular leader and the impact of the behaviors on the organization. We also prompt the assessors to not make other inferences about the person being assessed such as, "S/he has a lot of self-confidence." Instead, we would ask them for examples of what the employee actually does that gives others that impression. Having a standard, validated set of competencies also helps the discussion stay at the behavioral level and to avoid personality, inferred motives, or other references.

Calibrate, Calibrate, Calibrate

When you first ask leaders to conduct talent reviews, you are likely to find a wide range of interpretations of the criteria used to assess performance and potential. The agreement on the assessments can be significantly improved by having clear, behavior-based definitions and by using trained facilitators to guide the assessments. Conducting calibration sessions after the initial talent assessments are completed helps to ensure that the talent assessments are robust and reliable. Research has shown that handling the reviews as a series of structured discussions like this yields a complete and robust view of an organization's talent (Gandossy & Effron, 2003).

Calibration sessions comprise leaders from across units and functions who review a roll-up of the talent assessment results. The sessions are facilitated, and leaders are encouraged to challenge the talent assessment results. The challenges are encouraged if (1) the leader's experience with the employee is different from that reflected in the assessments or (2) leaders see inconsistencies among employees that are rated similarly. The same guidelines used in the talent reviews are followed in the calibration sessions—observations should be behavior based, objective, and reflect performance over time.

The goal of calibration is to reach a consensus on the picture of the organization's overall talent capability and to make sure that the process has identified the appropriate development strategy for similarly abled talent. The additional insights and comments about individual employees are captured and added to the talent database. Depending on the size of your organization, you will find that it is often beneficial to conduct separate calibration discussions by talent pools or levels.

Over time, you will find that participating in calibration discussions will improve leaders' ability to objectively assess talent and help make sure everyone is on the same page in applying the criteria. You will also find that some leaders are just not good at talent assessment and will need a lot of support and facilitation. Interestingly enough, we often find that leaders who struggle with assessment are not seen as high performers/high potentials themselves!

Performance and Potential Are Different

As we have emphasized, the goal of talent reviews is to identify the development strategies that are most effective and have the highest potential payoff for an individual or for pools of talent. That goal can be undermined if there is a tendency for those involved in the talent reviews to closely link Performance Effectiveness over Time and Growth Potential. This tendency had been labeled the "performance potential paradox" (Church & Waclawski, 2010). We emphasize in the preparation for and conduct of talent reviews that performance and potential are two very different constructs with different criteria. This is another place where the role of the facilitator is critical.

The More the Merrier

Multiple perspectives are critical. We've had the most success facilitating talent discussions with teams of three to five leaders that have been exposed to the target talent from a variety of perspectives. A facilitated, multi-rater approach avoids potential complications such as:

- Managers being overly positive about their own people (familiarity breeds non-objectivity)
- Leaders being overly or less candid about evaluating someone else's talent for fear that leader would retaliate in a similar fashion (tit-for-tat assessments)
- Political influence creating an intimidating environment (chest pounding)
- Hiding top talent for fear that a leader would lose them to another function (not in my house)

In addition to a multi-rater approach, calibrating the results at a high level across functions is the key to developing consistency and a consensus view of overall capability.

Don't Climb the Tower of Babel

All the players need to be speaking the same talent language. Leaders typically begin with different definitions of what should be covered in a talent review and how to define Performance Effectiveness and Growth Potential. (Just try asking a leader to define leadership!)

Therefore, everyone who participates in the talent assessments must be trained in the process and must use the same criteria and definitions for the quantitative assessments. The facilitators of the talent review process must also be rigorous in their emphasis of focusing on observed behavior over time and the impact of the behaviors. We provide participants with a reference sheet that identifies the key factors for discussion along with examples they can reference that help shape their comments. It's a great tool as we've had to tell even top leaders, "Make sure you check out your comments against our agreed-upon criteria." That usually stops the pontificating in a second. With experience, the managers and facilitators can become adept at converting the behaviors to standard competencies. It's easy to get people on the same page when you focus on observed behavior and when you have a common language (the competencies).

What Gets Measured Gets Improved

The output of effective talent reviews should be both qualitative and quantitative. The qualitative behavior observations are typically recorded to be used for feedback for Intentional Development. We tend to gather the qualitative, objective comments from the reviews in a database of some sort so that the information can be accessed for feedback purposes. Most modern Human Resource Information Systems (HRIS) have the capability to record and access this type of data.

Consensus and calibrated numeric ratings on Performance Effectiveness and Growth Potential should also be recorded. The ratings on each individual employee can then be rolled up to a total picture of talent that can be analyzed from a variety of perspectives and tracked over time to show progress. A simple spreadsheet and pivot tables can be used to slice-and-dice the data from a variety of perspectives and to answer a number of questions. In addition, most business analytics software or cubeware are an excellent and user-friendly way to analyze the data and compare results over time.

We have found that the much used, misused, and maligned "9 Box" to be an excellent tool to summarize the output of a talent assessment process. We say "misused and maligned" because there are many complaints in the literature about using a nine-celled matrix to summarize talent data. Most of the issues come from the fact that the data going into the matrix is unreliable and of low validity. Our approach to talent reviews— creating a unique tool we call the Development Strategies Matrix–directly addresses these concerns. More to come on this when we discuss the "9 Box on Steroids" in the next section.

Once Is Not Enough

Doing a talent assessment provides a snapshot of an organization's capability at one point in time. Multiple snapshots taken over time that are linked to business results significantly improve the utility of the process. We recommend that talent assessment at least be completed as part of the annual business planning process. Updated assessments and progress checks should be reviewed at least mid-year, if not quarterly. Repeating the reviews also captures the progress on development that occurs over time and defuses the practice of "placing someone in a box" as the purpose of the exercise. We have also found that leaders' assessment skills significantly improve with repetition and the sessions become more focused and efficient.

> "For a guy who hates bureaucracy and rails against it, the rigor of our people system is what brings this whole thing to life. There weren't enough hours in a day or a year to spend on people."
>
> Jack Welch, from his biography (Welch, 2001)

HR Is Not the Owner

Typically, talent management professionals are the conveners and recorders of the talent review process—but should not be the owner of the talent review process. While the ownership may ultimately reside at the top of the house, each leader needs to take the responsibility to track the progress of his or her employees and to use that information to best gauge their operational strengths and where their best opportunities for improvement lie.

All too often, we see that leaders think that talent reviews are for creating backup lists and, while this is a valuable exercise, being a strategy-driven organization requires that talent at all levels is being prepared to move on to new roles with either greater scope or scale . . . or both. This has been referred to as creating talent or acceleration pools (Byham, Smith, & Pease, 2015). This broader approach to talent assessment helps create a more agile, flexible organization that can better adapt to changing market conditions.

We would like to see Human Resources play a more critical role in the talent review process. The best role for HR in these situations is to go beyond convening and storing this data—bringing it alive and providing business leaders with the knowledge of how best to use their talent.

Mine the Data

The talent data gathered in reviews is very rich. HR should build the capability to effectively mine the data and provide proactive guidance to the organization. We will provide some examples of the analytics in a later chapter on talent analytics. In the absence of an integrated HRIS, we often create Excel databases to store and analyze the talent review data. We also have found that several Business Intelligence packages (such as Tableau, Power BI, or Qlik) are excellent in creating Talent Dashboards from the Excel databases or from the HRIS and other sources.

Take Action. Build Accountability

After the assessments have been gathered and calibrated, a critical last step in the talent review process is converting the talent data into a Strategic Talent Plan. This is where business strategy and talent come together. The plan should contain specific actions with assigned accountability, time frames, resources, and measures of success. We also emphasize that the plans should not be put on the shelf until the next round of talent reviews but should be part of the agenda of regular business meetings so that progress can be tracked, updates made, and corrective action taken, if needed. More on Strategic Talent Plans in the next section.

Here are some additional recommendations for implementing a robust talent review process.

- Based on the size of the organization, consider conducting talent reviews in cascading teams of leaders to cover several layers of the organization. Cascading talent reviews have the advantage of gathering insights on a larger number of people at deeper levels. As a result, the reviews often uncover people who have potential but have not been challenged or stretched.
- Avoid prescreening to exclude people from the review. Include all individuals at a particular level, unless they have not been in the role long enough to have a good sample of behaviors.
- Ensure the assessment discussions are led by a trained facilitator who guides the discussion with targeted questions, assures objectivity, and issues challenges when needed. If your organization has developed strategy-driven competencies, the facilitator may press for examples related to those skills.
- Provide advance training for everyone involved in the assessments. Ensure the training provides a clear set of criteria for the ratings and that everyone on the assessment team understands the criteria used to make final talent placement. Debrief each round of the process to look for continuous improvement opportunities and to provide feedback to the participants.

F. THE DEVELOPMENT STRATEGIES MATRIX OR THE NINE BOX ON STEROIDS

If you want one year of prosperity, grow grain,
if you want ten years of prosperity, grow trees,
if you want a hundred years of prosperity, grow people.

—Chinese Proverb

Once behavior-based evidence is gathered on each leader, assessors are asked to provide ratings on the two factors that are essential in determining effective development strategies—Performance Effectiveness over Time and Growth Potential. We described the logic and evidence behind these factors in the previous section. The assessors refer to a standard set of criteria when the ratings are completed and accompany the ratings with their logic. While we have used a variety of criteria and ratings over the years, Tables 4.2 and 4.3 describe what we have developed from experience and the research evidence.

By plotting the ratings on Performance Effectiveness against Growth Potential, we create a nine-cell matrix with each cell at the intersection of the different ratings prescribing a different development strategy, as you

TABLE 4.2

Performance Effectiveness Over Time Ratings and Criteria Example

Effectiveness Level	Contribution to Others' Success	Alignment	Accomplishments and Achievements
Exceptional	Consistently contributes to the success of others. Others react positively to how they achieve goals.	Always links goals and objectives to broader goals	Consistently among the highest performers. Consistently exceeds expectations.
Solid	Sometimes contributes to the success of others. Others sometimes react positively to how they achieve goals.	Sometimes links goals and objectives to broader goals.	Consistent performer that meets and sometimes exceeds expectations
Less Effective	Seldom contributes to the success of others. Often gets a negative reaction from others on how they achieve goals.	Seldom or never link their goals and objectives to broader goals.	Inconsistent performer. Seldom meets expectations

TABLE 4.3

Growth Potential Ratings and Criteria Example

Potential Level	Capability	Learning Agility	Aspiration
Agile	Demonstrates capability to handle larger and broader stretch assignments (more scope and scale). Could succeed in any situation.	Consistently demonstrates high levels of Learning Agility. Learns from experiences and applies the learning to new situations.	Consistently seeks out roles that would increase their contribution and development. Works beyond the scope of their job.
Expandable	Capable of growing into a larger role (more scope). Demonstrates ability to succeed outside their expertise. Experts that integrate their chosen area with other areas.	Demonstrates Learning Agility in some situations, or on some Agility dimensions, but not all.	Aspires to incremental levels of advancement
At Potential	Most likely successful in present or similar role. Experts in their chosen area. May not have yet demonstrated full capability	Seldom demonstrates Learning Agility or only in limited situations or on few dimensions	Satisfied at current level or has met overall aspiration

can see in Figure 4.3. Those of you who have been in talent management for some time will recognize this matrix as a significant reimagining of what's been called the 9 Box, the Performance Potential Matrix, the How Why Matrix and many other names. Beyond using the matrix to summarize the results of talent reviews in an accessible and easy-to-understand format, we've taken this venerable talent tool and significantly reimagined how it is created, how it is interpreted, how it is used and, as you will see, how it can be used as an analytical and tracking tool. We call our reimagined version the Development Strategies Matrix. Table 4.4 provides a comparison of the typical 9 Box process and the Development Strategies Matrix.

For our purpose, completing talent reviews and summarizing the results, in whatever format, has two outcomes:

1. Understanding the present capacity of your talent
2. Selecting the right development strategy for employees with similar performance and potential capabilities

TABLE 4.4

The Typical 9 Box Versus the Reimagined Development Strategies Matrix

Typical 9 Box	Reimagined Development Strategies Matrix
Used to classify talent—put employees in one of the boxes	Used to identify the most effective development strategies and allocating scarce development resources to the talent that will benefit from it the most. Movement among strategies is expected over time and across a career.
Business strategy plays little role	The whole process starts with the organization's strategy and ends with actions focused on the strategy
Criteria for performance and potential are not standardized	Uses evidence-based, behavior-based criteria on which all leaders are trained
Often decided by the input of just one manager or by a few people at the top of the organization	Multiple perspectives from those who are closest to the talent at all levels
Assessments are based on recent and highly flawed performance appraisals	Performance appraisal data is not considered. The criteria for performance considers multiple factors and looks over time
Often influenced by significant halo effect with most recent performance influencing all ratings	Trained facilitators lead the discussion and assure the discussion considers the individual in a variety of contexts over time.
Assessments and results are closely held by HR	Feedback to everyone included in the talent review is encouraged. Leaders are trained to provide the feedback to the person who was assessed.
No validation of the talent placement	Many of the development actions flowing from the process focus on confirming someone's potential through challenging assignments, 360 feedback, etc.
The focus is on the results	The focus is on the process and on how the resulting information can be used to drive change and progress.

Our focus, as we've emphasized all along, is not on "putting people in a box," but on development. Further, our focus is on strategy going forward, not on the past performance. Hence the name, Development Strategies Matrix. Figures 4.3 and 4.4 describe the 9 cells in the matrix.

The next section covers the next step in our strategic talent management process—establishing targeted actions to maintain and improve the leadership capability of your organization.

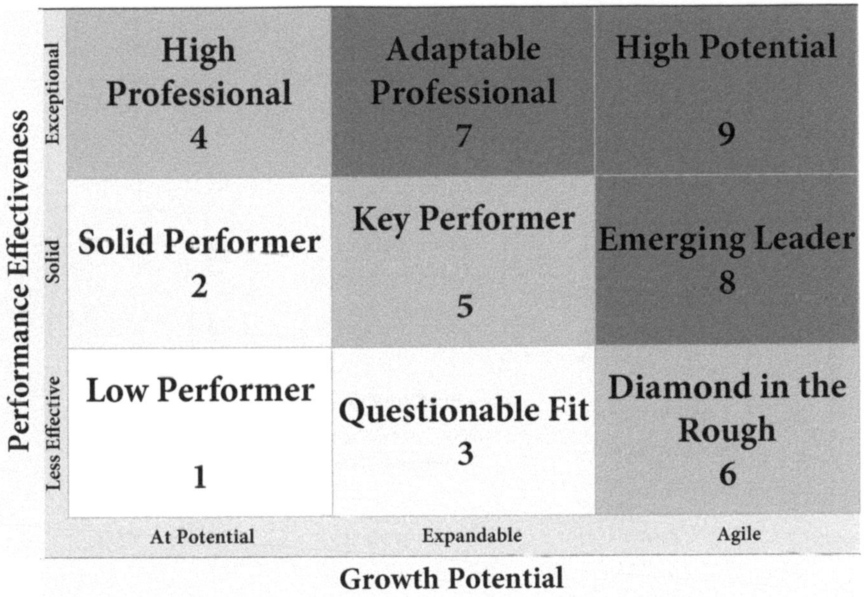

		High Professional 4	Adaptable Professional 7	High Potential 9
Performance Effectiveness	Exceptional			
	Solid	Solid Performer 2	Key Performer 5	Emerging Leader 8
	Less Effective	Low Performer 1	Questionable Fit 3	Diamond in the Rough 6
		At Potential	Expandable	Agile

Growth Potential

FIGURE 4.3
Development strategies matrix.

G. DEVELOPMENT STRATEGIES

You've completed the talent assessments and calibration sessions and have generated a picture of your organization's talent using Development Strategies Matrices. Now what? Well now is when the real development work begins.

No organization has unlimited resources to invest in their talent, so where do you start? To help us answer that question, we need to take a quick aside and discuss the idea of differentiated roles in an organization—what we call identifying strategy-critical roles.

A Note on Strategy-Critical Roles

As we mentioned previously, Becker, Huselid, and Beatty (2009) do an excellent job of describing how to identify strategic positions in an organization, the jobs that make the most difference. They described a rubric in

		At Potential	Expandable	Agile
Performance Effectiveness	**Exceptional**	**High Professional** Highly valued key contributor Invest and focus on Critical Knowledge Jobs	**Adaptable Professional** **Intentionally Develop** Actively develop potential for the Next Level of Leadership	**High Potential** **Intentionally Develop** Actively develop for Next Level Of Leadership
	Solid	**Solid Performer** Solid expert Continue Technical Roles	**Key Performer** Build Future Value Motivate and Reward Confirm Potential	**Emerging Leader** **Intentionally Develop** Actively develop performance for the Next Level of Leadership
	Less Effective	**Low Performer** Take Action and Fix	**Questionable Fit** Careful Next Assignment Engage and Motivate	**Diamond in the Rough** Potential performer Address Performance Monitor and Mentor
		At Potential	**Expandable**	**Agile**

Growth Potential

FIGURE 4.4
Development strategies.

which roles in an organization can be categorized as Strategic (A), Support (B), or Surplus (C).

- "A" positions have direct strategic impact and show a lot of variability in performance of people in those roles
- "B" positions have indirect strategic impact and exhibit little performance variation because they typically require following specific processes
- "C" positions are what we like to call business necessity; they are the roles needed for an organization to function, such as paying bills and paying employees

Our concern here is with clearly identifying the "A" positions with clients. We have found it helpful to break down the Becker et al. definition of "A" positions a little further and make it a little broader. In our view, "A" positions are strategic, pivotal, or scarce. Strategic positions are responsible for a key asset of the business (i.e., talent, finances, facilities, operations, or technology) and for setting the strategy for that chunk of the business. Such positions tend to be at the top of the house, and they have typically been the target of succession planning. These are the positions from which CEOs/presidents/executive directors are most often selected. They may also fulfill officer positions on boards.

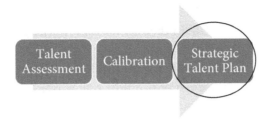

FIGURE 4.5
Talent review process highlighting the strategic talent plan.

Our delineation of "pivotal" roles probably fits most closely to the Becker et al. "A" positions. They have direct impact on revenue/market share, are typically in the core value streams of the business, and exhibit a large variability in people's ability to perform them effectively.

We have found that, in the current tight labor market, it is also valuable in talent assessment and planning to identify "scarce" roles. These are roles that require unique but critical competencies but that are also skills that are not common in the labor pool or are not easy to develop.

Strategic, pivotal, and scarce roles are not the same in every organization. As the name suggests, strategy-critical roles are derived from each organization's strategy. The strategy brings to light the roles that are vital to success, just as the key competencies are translated from strategy.

Okay, enough on differentiating roles. Now that the talent is assessed and the strategy-critical roles (strategic, pivotal, and scarce) are identified, it's time to develop a Strategy Talent Plan, as highlighted in Figure 4.5.

At a high level, here are the actions that we work through with organizations, in priority order of the payoff the action can create. Drum roll, please!

Strategic Talent Priorities

1. Remove Low and Inconsistent Performers (1, 3s) and possibly Solid Experts from strategy-critical roles
2. Move High Professionals, Adaptable Professionals, Emerging Leaders, and High Potentials (4, 7, 8, 9s) out from under low-performing or low-potential bosses
3. Intentionally Develop High Professionals, Adaptable Professionals, Emerging Leaders and High Potentials (4, 7, 8, 9s)
4. Look for opportunities to stretch Key Performers (5s) to see if they can actually handle more scope or scale

5. Make sure that Key Performers (5s) and Solid Expert Performers (2s) are staying up-to-date in their technical knowledge and are sustaining their networks
6. Mentor Diamonds-in-the-Rough (6s) to see if they can immediately improve their performance or find a role where they can perform or where you can confirm if they are truly Agile
7. Address the performance of Questionable Fits (3s) in non-strategy-critical roles
8. Address the performance of Low Performers (1s) in non-strategy-critical roles

Because we are writing about Intentional Development, you would think that we would start by focusing on that first and not as the third priority. However, we've found a greater payoff by first removing 1s, 3s, and maybe 2s from strategy-critical positions (strategic, pivotal, or scarce) and by making sure that your best and brightest are not reporting to leaders who do not engage and develop talent.

We feel that the drag on an organization's performance by having lower performing, lower potential people in strategy-critical roles is significantly underestimated. We worked for an organization that had never really fired an employee for other than cause . . . and the cause was seldom poor performance. The president was retiring so the board felt this was an opportunity to address this long-standing issue. We conducted a talent assessment which identified a number of low performers in critical roles. The president somewhat reluctantly took on the performance issues resulting in a number of high-level terminations. After it was all said and done, the normally staid leader called us into his office and effusively thanked us for our help. He was overwhelmed by the number of employees who came forward and thanked him for finally addressing the leaders who were holding everyone back. He said that it was the toughest thing he had done in his long career but it was also the best thing he had done. We've seen that same story play out many times since.

Intentional Development, then, is not the proverbial "kid with a hammer" to which everything looks like a nail. We feel it is more of a precision instrument that should be applied where it will have the greatest advantage—with the High Professionals, Adaptable Professionals, Emerging Leaders, and High Potentials and maybe some Key Performers. We've seen that the "one size fits all" approach is a recipe for waste, so a more differentiated approach is called for.

Table 4.5 provides an outline of the Developmental Strategies for each talent pool. We work step-by-step through the Strategic Talent Priorities and

TABLE 4.5

Development Strategy Descriptions

Development Pool	Development Strategies
Adaptable Professionals, Emerging Leaders and High Potentials (7, 8, 9s)	• Actively and intentionally develop for next level of leadership. • Identify lead roles within key stretch assignments to engage them in functioning at greater level of scope and/or scale and with a greater variety of other people • Place in pivotal, strategic, or scarce roles • Move them out from under lower potential leaders • Assign a senior leader who is a High Potential (or a High Performer in the case of an Emerging Leader) as a mentor • Expand their internal and external network. Expose them to Executive Leaders two or more levels above them and other High Potentials or High Professionals • Support Intentional Development Plans with an experienced executive coach
High Professionals (4s)	• Assign to critical knowledge jobs or projects; may be scarce roles • Provide opportunities for continuous learning; invest in maintaining and growing their expertise • Encourage them to develop their internal and external networks; expose them to other experts inside and outside the company and across functions • Recognize, reward, and leverage their expertise • Assign to leadership roles in their technical area • Move them out from under lower performing leaders. • Have them mentor/coach others to share their expert knowledge and to assure an effective transfer of knowledge over time
Key Performers (5s)	• Special assignments and projects at their current level. • Encourage self-directed growth and development • Check their capability, agility, and aspiration with temporary stretch assignments. Maybe test them at a higher level. Could they be a 7 or 8 with the right experience? • Recognize and reward their expertise • Leverage their unique knowledge across the organization • Build their internal and external network. Expose them to other High Professionals, mentors in their area of expertise, or others at their level across functions • Connect with others who would benefit from their expertise • If you want to test their growth potential, provide an executive coach who can work on their learning agility

(Continued)

TABLE 4.5

Development Strategy Descriptions (Continued)

Development Pool	Development Strategies
Solid Expert Performers (2s)	• Special assignments and projects at their current level • Encourage self-directed growth and development. Recognize and sustain their expertise • Make sure that they are staying up-to-date • Expose them to High Professionals, mentors in their area of expertise, or others at their level across functions
Diamonds in the Rough (6s)	• Urgently and candidly address problems with performance • Make a development plan with clear accountability • Provide opportunities to develop broader capabilities • Support with a leader that can coach or provide training to improve their performance levels • See if they want to change career paths. Are they misplaced? • If their learning agility is obvious, support with a coach.
Questionable Fit (3s)	• Urgently and candidly address problems with performance. • Remove them quickly particularly if they are in a pivotal or strategic role. May need to allow more time if in a scarce role. • Coach and train to improve key competency levels • Confirm fit with the role • Provide a combination of performance management and/or training to address specific competencies • Give a careful next assignment to increase performance and learning
Low Performers (1s)	• Take action: find a role that better suits them or remove them particularly if they are in a pivotal or strategic role. May need to allow more time if they occupy a scarce role. • Provide with clearly defined goals and clarify expectations about what they need to improve • Make little or no effort at retention or engagement. Seldom worth the effort • Work with HR to create a performance improvement plan and/or ethical, compassionate, and legal exit. (Remember the stayers)

refer to these Development Strategies to establish a Strategic Talent Plan with specific actions, accountabilities, time frames, and resources. In addition, we mine the data gathered in the talent reviews and refer to a variety of metrics to set baselines and to track progress. Those analytics are described in Chapter 8.

The Strategic Talent Plan should not be a one-and-done but should be reviewed as part of regular business review meetings. The analytics can also be updated to track progress or identify corrective actions if the plan gets off course.

5

The Solution: Intentional Development

A. WHY THIS WORKS: LEARNING THEORY AND INTENTIONAL DEVELOPMENT

Think for a moment about how your organization currently implements training and development. There are three major training delivery methods employed by US companies.

- **Stand-and-Deliver Training** from an instructor in a classroom. The Training Industry Report (2016) indicated that 41% of training falls into this category. Smaller and mid-size companies depend on this method more than larger firms.
- **Self-Paced or Online:** A little over 30% of US company training hours are delivered online or with other computer-based technologies. Virtual or Webcast delivery make up around 16%. Much of this instruction may be required as part of compliance programs (such as in health care or banking) or functional and technical knowledge unique to the firm.
- **Blended Learning:** About 28% of US companies employ a combination of training methods

FIGURE 5.1
Strategy-driven leadership development with emphasis on intentional leadership development.

In addition, a common talent process where development is *expected to* or *should* occur is within the annual *performance review*. It is in this process where a manager and employee identify the key areas of opportunity for an employee to focus on over the next year. The ideas may range from developing improved professional skills to mastering some aspect of time or meeting management, improving communication skills, or becoming a better writer. Typically written as a goal statement, the performance review objective is clearly enunciated but usually lacks operational guidelines other than to reference item some sort of training.

DOWN WITH PERFORMANCE REVIEWS

Experts now agree that the typical performance review process is terribly flawed. By trying to serve too many purposes (set goals, provide feedback, document performance, determine pay, etc.), it does none of them well. Survey after survey has shown that performance reviews are widely despised and perceived as ineffective. Brain research now confirms why. Just the act of *giving* feedback is perceived as a threat. Think about your reaction when the boss says, "Stop by my office. I have some feedback for you." The threat response is even worse when the performance discussion includes a performance rating. Your boss might as well have snuck up behind you on the savanna with a club! Likewise, HR's emphasis on documentation also creates a perceived threat—"Sign the form and it will be in your file!" Our human threat response is the same. We've consulted with many organizations who are moving away from their old appraisal processes and replacing them with regular, meaningful conversations . . . with no ratings.

In most situations, managers might review these objectives in a mid-year review, but with the lack of a solid plan and ongoing support, development work is slow at best and absent at worst. Either way, they generally miss the important mark of identifying what that employee needs to develop to bring more value to their work, which will discuss when we get to "Frame It."

The failings of these approaches miss out on several of the key factors that are related to how adults learn. Developing and keeping a growth mindset about adult learning will significantly improve the investment in the effectiveness of leadership development. Let's examine some of the cornerstones of adult learning strategies.

Adult learning strategies are not one-sided. It is not just the role of the instructor or manager to shape the learning experience. The student must be a participant as well and interestingly, most adult learners enjoy learning more as adults than they ever did as children.

Malcolm Knowles (1984), who was a pioneer in the field of adult learning, identified six key factors critical for success of adult learning. These include:

1. **Adult learning comes from real life experiences**. Not only do adults learn through the experiences they've had but even more importantly, they learn from new experiences that they face. Putting your people into real life situations where they can develop and refine skills is one of the best ways to make learning come alive.

2. **Build buy-in through self-direction**. One of the key challenges with performance review–driven development planning is that it is usually developed by the manager. The employee is usually asked for a self-evaluation but most reports indicate that the employee's side of the equation is grounded in trying to protect too negative a report. Furthermore, the employee perspective usually lacks input into the direction for the growth. A more effective learning approach is for managers to serve as mentors or coaches to help employees reach the goals they want to achieve. This helps create the buy-in on the part of the employee that is a critical aspect for their taking ownership of their learning.

3. **Make it relevant**. Remember when you were in ninth grade and asked whether you would ever be using algebra? It's a good thing your mother came up with some good examples to keep you motivated. Unfortunately, that approach does not work for adult learning and so managers must use the Frame It perspective to help employees keep their focus on business strategy and outcomes that ensure that development-related experiences are tied to real world situations.

4. **Make it meaningful or goal-oriented**. Along with relevancy is the notion that the learning experience is going to help the employees accomplish something meaningful in their lives. Creating a learning

plan (or what we call a Learning Map) establishes specific objectives for the employees to use that provides objectives, keys to understanding, strategies for learning, action steps, and building in sustainability, all of which provide fuel for learning acceleration and takeoff.

5. **Collaborate with others**. Learning becomes more fun when it is done is collaboration and, in an environment, where tasks are shared and learning ideas can be reflected upon with others. Most of our best clients establish "cohort groups" that provide each participant with a forum group for examining meaningful corporate projects along with accountability partners that create a strong environment for success and for an exchange of ideas between people.

6. **Support learning with feedback**. Adults are smart and, given information, they will shape and change their actions. We emphasize four key factors about feedback in Intentional Development: it must be relevant, frequent, a mix of positive and negative and *requested* not imposed feedback. There is danger in providing unsolicited feedback. Think about how you respond when someone comes over to you and says, "I have some feedback for you; I know you didn't ask for it, but I'd like to give it to you anyway."

LEARNING AS AN ADVENTURE

Recently one of us (Richard) was talking to a coaching client who is a senior executive in a pharmaceutical firm. She was with the company more than 22 years and had progressed to that high level through her smarts and hard work. But she had a little secret she kept to herself which was that she never graduated college. She had taken classes as a young woman, but family issues kept her from finishing her coursework. She told the story of how she always wanted to finish college and recently realized that although she was very successful in her present role, if she needed to change companies, her lack of a degree could be a significant impediment to her career. She went back to school with a commitment to finish quickly and with good enough grades to move on to graduate school. At the time we started working together, she was managing three classes, her new role as a senior leader, and being a mom and wife to two teenage sons and her husband.

When I asked her how she was able to do all this, she said that she loved school and was looking forward to graduating in the next semester after taking three more classes (and these were classroom programs, not online programs). Learning, for her, was an adventure and was not something that was an "extra to do" but was an important, valuable, and worthwhile addition to her life.

The purpose of any development effort is to help leaders acquire, change, or enhance critical competencies—to do more of something, to do something new, to do it better, or to do it differently. In a chapter of his book on coaching, David Peterson (2006) provides an excellent summary of the research that defines the "necessary and sufficient conditions" that must be created in a learner for change to occur—conditions which he calls the "Development Pipeline." The necessary conditions are:

1. Insight: The extent to which the learner knows what he or she needs to develop
2. Motivation: The extent to which the learner is willing to invest the time and energy to change
3. Capabilities: The extent to which the learner has the skills and knowledge needed
4. Real-world Practice: The extent to which the learner receives and uses opportunities to try the new skills on the job
5. Accountability: The extent to which the learner internalizes the new skill and sees meaningful consequences from applying it

This foundation work on general factors for successful adult learning and behavior change were great places for us to start in building our approach to intentional employee development. We've taken what we've learned in the ensuing years and additional research to identify 10 essential tools that form the foundation of Intentional Development which are outlined in the next section.

We also wanted to create a simple, effective process that encapsulated all we have learned about development and that could be applied across an organization. We came up with *Frame It, See It, Do It, Connect It* to outline our approach to Intentional Leadership Development. We will delve into the "*It Formula*" in the next chapter.

B. INTENTIONAL LEADERSHIP DEVELOPMENT ESSENTIALS

Here then are the 10 essential tools that form the foundation of Intentional Leadership Development. We'll describe each one in more detail.

1. Have a Planned and Targeted Impact
2. Focus on the Critical Few Competencies

3. Build It In; Don't Bolt It On
4. Reject "One Size Fits All"—Mass Personalization
5. Understand There's More Than One Path to Development
6. Create a Cadence of Development
7. Create a Feedback-Rich Environment
8. Match the Development Strategy to the Talent
9. Make the Development "Sticky"
10. Never Learn Alone

1. Have a Planned and Targeted Impact for Development Initiatives

Intentional Development begins by framing the development in a way that clearly links outcomes to effort, linking "Why is this important?" (the link to the business strategy) to "How is this accomplished?" (an Intentional Development Plan). Setting the context in this fashion not only capitalizes on two of Knowles's adult learning factors (making it relevant, making it meaningful); this framing also helps our brains be prepared for and pay attention to something new. As described in the article, "Learning that Lasts Through the Ages" (Davachi, Kiefer, Rock, & Rock, 2013), certain neurochemicals need to be released in our brains to create a state where we are curious and open to gain something new. The levels of those chemicals increase when we see value in a situation and when we see the situation as relevant to us personally.

The number one factor identified in the McKinsey summary of why leadership programs fail (Gurdjian, Halbeisen, & Lane, 2014) was "over-looking context." As the authors state, "Too many training initiatives we come across rest on the assumption that one size fits all and that the same group of skills or style of leadership is appropriate regardless of strategy, organizational culture, or CEO mandate. In the earliest stages of planning a leadership initiative, companies should ask themselves a simple question: What, precisely, is this program for?" (p. 2). Context matters in leadership development and is the place that all development must start.

Having a planned and targeted impact for development does more than prepare individuals to learn. It puts employee development on par with other business functions. In multiple ways all organizations focus on the planning. For example, budget and operational plans, logistical project planning, and IT implementation plans all meet a standard of establishing a starting point from which projects begin. Traditionally, professional development for employees has not met the same standard as these other business functions. While we often hear that "there is just not enough time," that explanation does not work for budgetary or operational planning. These efforts might consume the entire last quarter of the fiscal year.

Tying development to business planning and outcomes is a certain way to increase the perceived value of development as a business outcome, along with helping to improve actual business outcomes.

2. Focus on the Critical Few Competencies

We came across an example of a competency model that a healthcare organization had created for their leaders that contained 25 individual competencies. When we asked the HR leader how that was working for the hospital, she indicated that it was perceived as cumbersome and not well understood across the organization.

Intentional Development clearly identifies the skills required to improve a learner's performance or increase his/her potential for future success. We like to use the phrase, "You can't be it until you see it" when we talk about this stage of the process. The "it" needs to be a few basic, simple development themes, not a long, exhausting list of skills. Customizing and personalizing development in this fashion has been shown by others (i.e., Jensen, 2005) to be a key in optimizing learning and retention. The fewer competencies that are included in an Intentional Development Plan, the more likely the learner will make progress and have an impact with new skills.

Lou Gerstner Jr, the chairman and CEO of IBM from 1993 to 2002, describes in his 2002 biography how IBM used competencies to revitalize the ailing company. He felt that the competencies did help create more effective leadership behaviors and a new focus but thought that even the single set of 11 competencies they developed was a little too complex. To address that concern and to make the competencies more memorable, they were clustered into three strategic themes: Win, Execute, and Team (Conger, 2010.)

One way to customize development with individual learners is to employ the concept of differentiating strengths espoused by our innovative colleague, Tracy Fuller. Tracy highlights that, "Yes, it's important to know your leadership strengths. But that's not enough. To excel at a given time and in a given situation, you need to know your *differentiating strengths*" (Fuller, 2019, p. 1). This concept, as diagrammed in Figure 5.2, emphasizes

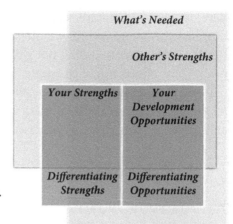

FIGURE 5.2
Differentiating strengths and opportunities.

Source: Adapted from Fuller (2019) Used by permission of the author.

that knowing strategy-critical competencies (what's needed) is important. However, understanding the context of a developmental challenge along with awareness of one's strengths and development opportunities versus others may help target competencies that are different than what would normally be emphasized. We've added to the approach by suggesting that learners can think about development opportunities in the same way—as differentiating. In addition, conceptualizing competencies as differentiators in this fashion helps individuals envision what might better constitute their personal leadership brand.

3. Build It In; Don't Bolt It On

We often begin Leadership Development discussions by asking the question, "What occurred in your past that makes a difference in how you lead today?" We've probably asked this 100 times and the results are always the same. Leaders report that they learned meaningful skills by navigating a variety of challenging experiences. Coupled with these developmental experiences were other people that provided developmental support, feedback, or mentoring, along with just-in-time study, training, or research.

The pattern of experience, people, and study that we heard matches the research conducted at the Center for Creative Leadership and subsequent global studies (Robinson & Wick, 1992; McCall et al., 1988, McCall & Hollenbeck, 2002, Fernandez-Araoz, 2014). This research identified that effective development has three key components that follow a 70–20–10 ratio: navigating challenging roles or assignments (70%), developmental support, coaching, mentoring, or role-modeling from others (20%), and self-study or training (10%).

A quick aside on 70–20–10: other studies have found different percentages for the three components (Sinar, Wellins, Ray, Abel, & Neal, 2014) but always in the same relative order. We have also found some client companies whose leaders report a historic pattern that also varies from the exact 70–20–10 ratio, usually with more emphasis on the "learning from others" component. The exact numbers aren't the issue. The point is that developing critical leadership skills involves a combination of these three elements with the emphasis placed on navigating challenging experiences. For simplicity's sake, we will use the 70–20–10 nomenclature through the rest of this book.

For development to be effective, it must be built into a leader's day-to-day work and not be seen as something extra that has to be added to an already busy schedule. The only corollary to Build It In occurs when the present job is not challenging enough. In this case, the leader's role needs to change, the job needs to be revised, or more challenging assignments need to be added.

Research has found that how challenging an experience is predicts how much learning will occur as a result of the experience (McCauley, Ruderman, Ohlott, & Morrow, 1994). Without significant challenge (or what Lombardo and Eichinger, 2000 referred to as "Developmental Heat"), our brains remain on autopilot and employ the behaviors that have been stored in long-term memory. As the neuroscience research has highlighted, human brains are pattern recognition machines that like certainty. Only when patterns are perceived as significantly different do our brains snap out of standard operating procedure and kick in their energy-intensive, short-term memory prefrontal cortex. Focused energy and attention creates new neural connections that are embedded in long-term memory. These new connections/habits/behaviors can then be easily recalled when needed—sometimes unconsciously.

Most key leadership challenges involve plenty of potentially brain-changing opportunities. Assuring that learners focus their energy and attention during those learning opportunities is a hallmark of Intentional Development. What might some of these challenges look like? Table 5.1 summarizes the research (McCauley et al., 1994; Yip & Wilson, 2010) on the types of challenges that have been shown to be the most developmental.

TABLE 5.1

Key Developmental Challenges

Developmental Challenges	Description	Examples
Taking on unfamiliar responsibilities	Proving yourself by handling responsibilities that are new, very different, or much broader than before with more decision-making power, influence, and visible opportunities for success or failure. New knowledge and expertise are needed that the leader does not have and previous knowledge and experience may be inadequate	• Experience a major change in one's work role/position • Manage something with which you are unfamiliar • Transfers of all kinds—line to staff, staff to line, new geographic region, division to headquarters
Creating change	Creating and facilitating change in the way business is conducted or fix a preexisting problem in conditions that are neither clear nor predictable	• Starting something new, making strategic changes, carrying out a reorganization, or responding to rapid changes in the business environment • Turning around or fixing problems created by the former incumbent • Making decisions about shutting down operations or staff reductions that have to be made • Managing a new product launch or acquisition • Handling employees who lack adequate experience, are incompetent, or are resistant
Assuming higher levels of responsibility	Leading initiatives that are highly important to the organization and entail multiple functions, groups, or products/services	• Leading assignments with clear deadlines, pressure from senior management, high visibility, and responsibility for key decisions make success or failure in this job clearly evident • Taking on a role with a large scope such as with responsibilities for multiple functions, groups, products, customers, or markets • Dealing with external factors that impact the business (e.g., negotiating with unions or government agencies, working in a foreign culture, or coping with serious community problems)

Developmental Challenges	Description	Examples
Working across boundaries	Influencing or managing people or processes for which one has no direct authority. Reconciling different points of view.	• Influencing peers, higher management, external parties, or other key people over whom the manager has no direct authority • Convincing upper management to support a proposal • Managing key interactions with an important labor union
Working in a different culture	Leading people from different cultures, gender, leadership practices, racial or ethnic backgrounds and/or languages. Managing diversity	• Leading a team dispersed across several countries • Leading a team with extensive gender and racial diversity • Taking an expatriate assignment

BUILD IT IN, DON'T BOLT IT ON

"I don't have time for this," is the lament most often heard by facilitators of professional development programs. Employees already feel overwhelmed by their consistently full to-do lists, so attending a training session creates a challenge many of them do not want. We tell our coaching clients that if they feel that our approach to development is creating more work for them instead of less, then we have failed.

In our model, professional development is "built into" the employee's day-to-day activities instead of being "bolted on" as an additional task that has to be completed like another project. Our development model promotes learning new skills and competencies in the context of the person's regular job. And we believe that the skills the individual is learning will help the person perform regular duties more efficiently.

Recently, we were working with a company that asked us to assist the director of Corporate Safety to improve his presentation style, which was bland and monotonically presented. He had taken some classes and even spoken to a speech coach. The approaches had him practicing new speeches and presentations that he had to develop, refine, and then test in front of other audiences. It seemed obvious to us he had plenty of opportunities within the context of his role to practice the relevant skills. There was no need to create artificial situations. We helped him review his slides, read his content, practice in front of us, and then sent him off to present in front of the senior leadership group. It took just a couple of months of working with him to get his technique refined, but the great advantage of this approach was that his managers could see his incremental improvement and, with their support, his presentations improved consistently.

4. Reject "One Size Fits All"—Mass Personalization

It is highly unlikely that members of a given set of leaders have the same developmental needs. Why? Because no two leaders have been through the same developmental experiences or have the same ability to learn from experience. That's why sending all leaders to the same leadership development workshop is seldom effective.

One of our client organizations decided they were going to send two of their high-potential employees to get an MBA from a local university. One of these staff members was a senior level operational leader, who was also an up-and-coming strategic leader. The other served as the controller in the finance department. While highly gifted in his space, he, at that point, had failed to demonstrate an ability to think beyond finance.

The CEO hoped that sending both leaders to the same program would provide a learning forum for them both. As you might imagine, over the course of the program, the ops leader strengthened his skills and began to contribute to the larger strategic mission of the organization in a much more meaningful way. The controller continued on his focused path, despite attending classes designed to broaden his horizons.

It probably would have been better to send the controller to a certification program that focused on helping finance people see the strategic role they could play within an organization rather than a full MBA program. Even though both leaders were identified as having the same needs in terms of how to benefit the organization, only one of them gained what the organization needed.

Developmental needs differ across learners, but the process that leaders must maneuver to create new competencies is the same—Intentional Leadership Development.

5. Understand There's More Than One Path to Development

When receiving feedback, all of us naturally focus first on perceived weaknesses—we tend to have what Richard cites in his book *The Resilience Advantage* a "negativity bias" (Citrin and Weiss 2016). It is a tendency that is hardwired into each of us that says we need to make sure we pay attention to possible risk factors that might impact our lives in an unhelpful or even dangerous way. Improving a weak skill can have its benefits. However, other development paths can be more beneficial or make a greater difference.

For example, results of Center for Creative Leadership's Lessons of Experience (McCall et al., 1988) research demonstrated that leaders often failed in their careers by *overusing* an existing strength rather than because they had a glaring skill deficiency. In that case, reducing the overdependence on strength could be the most effective and highest impact development strategy. Eichinger, Lombardo, Stiber, and Orr (2011) in their book, *Paths to Improvement*, describe 14 potential approaches to developing leadership competencies grouped into the following five categories.

- Deeper Exploration—gaining more insight into the skill or trying out the skill in a new experience
- Direct Skill Building—working on enhancing an okay skill to a strength or moving a strength to exceptional
- Alternative Paths—substituting, working around or compensating for one skill with another, particularly if the skill is tough to develop
- Demonstrating the Skill—marketing the skill to others, transferring a skill to a new situation, and building confidence in the skill
- Accepting the Consequences—not actually developing the skill but using other strategies such as selecting others for your team that have the skill or by moving to a role that better matches your skills

To create an effective Intentional Development Plan, we consider:

- The importance of a skill (as highlighted in the organization's strategy and an Impact Map)
- The type and availability of experiences needed to build the skill (the 70 of the 70–20–10 model)
- The ability and willingness of the learner to envision the new skill and its benefits
- How difficult the targeted competency may be to develop or how many different experiences it may take to build competence

6. Create a Cadence of Development

Developing a new key leadership skill requires regular focus and attention—what we call a cadence of development. Effective development is never a one-and-done. That's why event-based training is a dead-end when it comes to building leadership skills. Intentional Development is just that—intentional on both sides of the equation. The learner has to think about the new skill and the situations in which he or she will apply it. And the learner needs to focus on it regularly as it relates to what he or she wants to do more of, differently, or better.

We are not referring to the old saw that says you need to repeat something a set number of times (usually large) to learn it or for a new behavior to become a habit. In fact, research has shown that repetition has very little impact on learning that sticks (Davachi et al., 2013). To increase the likelihood that we will retain and recall something new when needed, we need to reflect on learning experiences and match them to our previous experience or knowledge (Jensen, 2005).

We ask our learner executives to reflect on and record their Intentional Development efforts at least weekly. We ask them to report what new behavior they tried in what situation; what worked or didn't work; what the impact was of the effort; and what did they learn that they will apply in similar situations in the future. This approach mimics a type of micro-learning methodology where small, incremental, and regular learning opportunities reinforce the larger learning objective and where this regular reflection helps to lock in the newly acquired behavior.

7. Create a Feedback-Rich Environment

The basic overarching operating principle of human brains is to maximize reward and minimize danger or threat. But our brains are much more attuned to detecting threats—it is much easier to cause aggravation and avoidance in social situations than it is to generate positive emotions. (There's that negativity bias again!) As a result, even the idea of getting feedback is often perceived as a threat that is to be avoided (Rock & Ringleb, 2013). As we mentioned earlier, you just need to think of your own reaction if your boss called you into her office and said, "I have some feedback for you."

The challenge in creating a feedback-rich environment, then, is to reduce the threat-avoidance response. It turns out the best way to do this is to have more frequent, two-way discussions between managers and employees and to encourage learners to ask for feedback rather than to have it imposed. Research by Deloitte has shown that 90% of companies that implemented processes for continuous feedback from managers saw increases in employee engagement (Sloan, Agarwal, Garr, & Pataskia, 2017).

Objective, targeted, and *requested* feedback plays an important role in the See It, Do It and Connect It stages of Intentional Development. We plan for and gather feedback in a manner that increases the likelihood that the information is perceived as rewarding and beneficial. Even negative feedback can be seen as a reward if it's collected in the context of an Intentional Development plan that is framed with clear benefits for the learner.

While it is clear that navigating challenges is the best way to develop new leadership competencies, research has also shown that the developmental value of experiences can begin to diminish at a certain point, especially for the development of interpersonal and business leadership skills (DeRue & Wellman, 2009). However, the same research found that the individuals who had access to feedback were less likely to see this diminishing effect, as described in Figure 5.3 Further evidence of how important the 20 is in the 70–20–10 model of development.

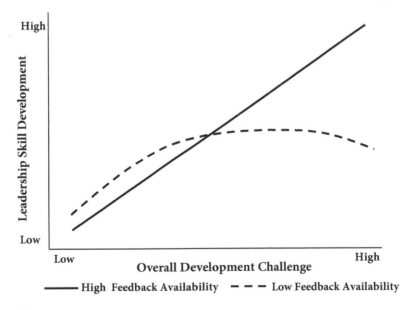

FIGURE 5.3
Moderating Effect of Feedback Availability on the Relationship Between Overall Developmental Challenge and Leadership Skill Development (DuRue & Wellman, 2009).

Research has also shown that providing more regular, ongoing feedback and more developmental feedback were strong predictors of financial performance including net profit margin, return on investment, return on assets, and return on equity (Ledford & Schneider, 2018). The same study found that the impact can be enhanced by building a "feedback culture" through practices such as better communications on the importance of feedback, training on how to provide effective feedback, monitoring the effectiveness of feedback, and even rewarding and selecting managers that were effective at giving feedback.

8. Match the Development Strategy to the Talent

As we will highlight in Chapter 6, not all employees will benefit from the same development strategy based on their current performance, potential, and aspiration. This is where understanding the organization's and individual leader's present capability is essential. (See the right side of Strategy-Driven Organization Development model.)

Since no organization has unlimited resources, it is important for you to know which development strategies will have the greatest return on the time and money invested. For example, placing High-Performing/High-Potential employees in challenging, pivotal roles has benefits for both the company (productivity and growth) and the employee (engagement and rewards). However, applying the same development strategy for employees that aspire to be technical experts in the company's core competencies would be a disaster. The technical employees would disengage and the company would lose the benefit of their unique knowledge. A high-impact development strategy for the technical experts would be to assure they are focused on maintaining their expertise, thinking about the application of the knowledge for continued competitive advantage, and transferring their knowledge to other solid performers that could be potential technical experts.

9. Make Development "Sticky"

Research summarized in the *Handbook of NeuroLeadership* describes how envisioning a behavior can be equivalent to actually performing the behavior (Dixon, Rock, & Ochsner, 2013). Imagining activates the same areas of the brain as actually performing the behavior. Mental rehearsal of a behavior can become an embedded memory that can be recalled—it makes the development what David Rock of the NeuroLeadership Institute referred to as "sticky."

This is akin to the benefit that sports psychologists discovered from having athletes mentally rehearse the perfect golf shot or the record pole vault before actually attempting the feat. Rehearsing the intention also improves the chances that the learner will actually take some action and learn a new skill.

For example, what new skill do you think the folks in Image 5.1 are attempting to develop?

This is Sweden, September 3, 1967, the day the country switched from driving on the left to driving on the right, known as Hogertrafikomlaggningen ("the right-hand traffic diversion") or H Day. If we counted correctly, it looks like only five cars got it right at the time the picture was taken. Everyone else just gave up and went back to an embedded behavior that they knew worked—walking.

Think about the mental process you have to go through to build the new skill of driving on a different side of the road. You may have gone through this if you ever had to drive in a country that drives on the "wrong" side. You have to be intently focused and pay attention. You have to think to yourself, "I'm approaching a roundabout; I need to go around in a different direction or I'm going to have a head-on collision" or "I'm at an

IMAGE 5.1
Hogertrafikomlaggningen.

intersection; to turn left, I need to pull into the lane closest to me, not the farther lane or I'll be going the wrong way." It's mentally exhausting and somewhat stressful until you build the new habits. From a neuroscience perspective, you are using your energy-intensive, short-term, limited-capacity memory to create new connections in long-term, unlimited-capacity memory that can be easily recalled when needed.

As part of their development planning, we ask learners to come up with Intentional Development "mantras" or "rules of thumb." These are short statements in the form of "If I find myself in this situation, I am going to do this, so that I achieve such and such an outcome"—just like the mental maps you use to make sure you drive on the correct side of the road. We encourage learners to refer to these "if then, so that" statements regularly to create the focus and attention needed to build new sets of behaviors in long-term memory.

For example, we had an executive client who was seen as rather unapproachable and tough to get to know. To address this negative perception, he created and employed the simple mantra, "When I meet an employee for the first time or see one of my folks in the hall, I'll make sure to smile, say hello, and ask how he or she is doing so that people see me as more friendly." Simple, but for him, it turned out to be very effective; people began to quickly increase their interactions with him.

In another case, a marketing analytics professional's multi-rater feedback showed that she was perceived as being an ineffective communicator and lacked a strategic perspective. Her reports were technically sound, but her clients didn't know how to incorporate the analytics into their decision making. Her mantra became, "When completing an analysis, I will always provide clear recommendations so that others see my communications as more effective and linked to our strategy." In the first opportunity to apply her mantra (and change her behavior) her recommendations were taken seriously and immediately implemented. That hadn't happened before.

Our intentional use of mantras or rules of thumb in development builds off the neuroscience associated with the concept of "implementation intentions" (Gollwitzer, 1999). The basic idea is that an individual will more likely achieve a difficult goal or learn a new skill if they specifically prepare in advance what to do in a particular situation. Swedish psychiatrist and neuroscientist David Ingvar described this unique capacity of the human brain as the ability to create "memories of the future" (Ingvar, 1985). We've adapted the neuroscientist's suggestions on implementation intention statements to outline four possible intention strategies that help

TABLE 5.2

Intentional Development Mantra Examples

Intention Strategy	If . . .	Then . . .	So that . . .
Change the Situation	I am asked to be a member of a project team	I am going to request that I lead the team	I am seen as taking more initiative and benefit from playing a different role
Modify the Situation	When I am asked to do marketing analyses	I'm going to include my own observations and recommendations	I am seen as being more strategic and as having a broader understanding of our business
Change Your Focus	I need to assign work projects to my team	I am going to consider more than who is available and think about who might benefit the most from the experience	I help my team members develop new skills and become more engaged
Change Your Mind	I find myself getting upset at Mary's outbursts	I am just going to listen and take notes	I reduce the conflict between us and not waste so much time worrying about it

learners build future memories and make development sticky. The four strategies and example mantras are summarized in Table 5.2.

The "ifs" are key situations where improved performance would make a difference in the learner's or his/her team's performance. The "thens" are linked directly to targeted competencies that the learner would like to try out, do more of, do differently, or do better and are most helpful if they are specific actions/behaviors. The "so thats" are the benefits or impacts that the learner expects to see from applying the competency. These are all highlighted in their Impact Maps, Learning Maps, and Intentional Development Action Plans. More on those in later chapters.

We encourage learners to refer to their mantras regularly, even daily. That way they can keep their Intentional Development front-of-mind, focus their attention, and stay alert to opportunities to practice their target development competencies.

10. Never Learn Alone

Leaders of a large retail client wanted to determine how to sustain or enhance nine components of their culture to drive growth. We worked with them to assemble nine cohorts of leaders for whom this would be a great developmental challenge. The teams were assigned two tasks: (1) research the culture factor and make strategic recommendations on how to leverage the factor to drive growth and (2) use the project to become better leaders by building new skills. After the teams reported their results to the executive team of the company, the top executives commented that the recommendations were great—but the increase in the capability of the leaders may have been the bigger payoff from the project. The cohort members reported the same, particularly the benefit they had gained by working with their cohort members.

Humans are intensely social beings. We are just starting to understand the benefits of social networks within organizations. One benefit, reported in research by CEB (2013), is that the extent to which people not only do their own jobs well but *contribute to the success of others* is related to overall performance of organizations. This is why including network contribution in the assessment of talent is critical, as we mentioned previously.

There seems to be a neurological basis for our social bias. Research has confirmed that, as David Rock has stated, "The human brain is a social organ. Its physiological and neurological reactions are directly and profoundly shaped by social interaction. . . . The brain experiences the workplace first and foremost as a social system" (Rock, 2009, p. 3).

We leverage this social bias and social networks by employing Development Cohorts as a regular component of Intentional Development. Cohorts provide a safe environment for building new skills and significantly enhance the 20 of the 70–20–10 model of development. The cohorts can focus solely on development; however, the learning opportunity is significantly enhanced when the cohort is also assigned a meaningful challenge to address. This combines a 70 with a 20, adding in the Intentional Development factor of "Build It In; Don't Bolt It On".

We find that the employing Development Cohorts helps learners establish a trusted network of peers that is an additional source of objective feedback outside of their normal feedback channels. We also encourage the members of cohorts to select "Accountability Partners" with whom they can review progress on their Intentional Development Plans and commit to specific follow-up on the plan.

6

The Intentional Development Process

A. INTRODUCTION

When working with client companies around the time of the financial meltdown, we used to tell them that now was the time when employers would have the upper hand in finding and developing talent. Employees were desperate to have a job they could hold on to and they would do their best to perform for the organization. However, we warned our clients that this time would pass. Unless they used this time wisely to build a talent structure that grew and supported their development, they might find themselves on the other side of the equation and trying to catch up to finding good people when those people all started moving to other firms. Few leaders followed our suggestion, but we had one client CEO emphasize to his leadership team that he wanted to come out stronger on the other side of the recession . . . so now was a time to focus on developing talent. It was his unique version of "Why waste a good recession?"

Today, however, the situation is completely opposite. With unemployment at record lows and generational shifts that include significant Baby Boomer retirements and greater expectations among Gen X, Y, and Zs around workplace satisfaction, employers are dancing as fast as they can to keep their talent pools intact and growing. This younger generation of workers sees through shallow efforts or "all talk" ideas about their opportunities for growth and development and don't want to wait until it is "their turn." They want to demonstrate that they can make a meaningful contribution. Boomers want to continue to find meaning and generate increasing compensation to fund what might be a long retirement, and so they are interested in new opportunities that will help fuel their success and cement a personal legacy. The result is that employers need to be intentional in finding a way to create meaningful learning opportunities for all employees or face the potential of becoming a talent desert.

DEVELOPMENT AT BLACKROCK

Blackrock is the world's largest financial asset management company. While their success can be attributed to many factors, including their financial and technological acumen, the leaders readily acknowledge that their commitment to developing a world-class talent organization is central to their mission. In the world of financial management, the difference between success and failure relates to how well the leadership and talent of the organization execute every day.

Like so much of the work that is done at Blackrock, their focus on talent development is nearly maniacal. As written about by Douglas Ready and his colleagues in a 2014 issue of the *Harvard Business Review*, Blackrock's Human Capital Committee, composed of 35 senior leaders (the only HR professional is Blackrock's Global Head of HR), oversees their approach to talent. Several tactical steps are important in creating actions that can be taken such as talent reviews, succession planning, networking, and collaboration tools—but the bigger themes of driving high performance culture, prioritizing behaviors that matter, and being intentional about how they develop their people create an underlying value system that supports their overall mission of service to clients.

B. THE INTENTIONAL DEVELOPMENT "IT" FORMULA: FRAME IT, SEE IT, OWN IT, CONNECT IT

No organization can afford to ignore the need for talent development. We believe that a structured, transparent approach to developing talent works best when done in the open where everyone, from senior leader to line worker, knows what the expectations are to become an accomplished and successful leader in their company. We describe a four-step process for accomplishing this task in an intentional manner. See Figure 6.1.

Frame It: As we've described, tying leadership development to the key business strategy and/or key personal outcomes is essential. Resources are too sparse to fail to target a specific goal or objective critical to the firm's success into any development activity. Context is key.

See It: We're all about behaviors, and you should be too. While, as psychologists, we might be interested in what makes people tick or why they

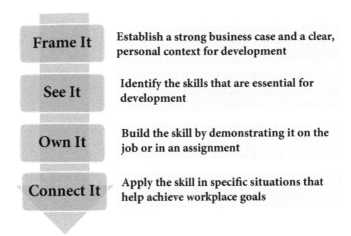

Frame It   Establish a strong business case and a clear, personal context for development

See It   Identify the skills that are essential for development

Own It   Build the skill by demonstrating it on the job or in an assignment

Connect It   Apply the skill in specific situations that help achieve workplace goals

FIGURE 6.1
Four phases of intentional development: the it formula.

behave in a certain manner, our greater focus is on ensuring that people understand the behaviors that build success for themselves and others. There is no more powerful path to leadership success than self-awareness and understanding what it is that you do well and where you need to improve.

Own It: Development comes from navigating a challenging experience and learning or enhancing strategy-critical competencies from the experience. During Own It, learners create maps and plans to chart experiences which require them to deploy new behaviors (or to learn the right stuff from existing challenges) and to lock-in-the-learning from the experiences.

Connect It: Connect It focuses on assisting learners to effectively apply new skills in real-life business situations. In addition, this step helps leaders to understand how they can apply Intentional Development concepts and tools across a career. Connect It concentrates on assuring that key developmental experiences are positive (additive) to a career and that the stage is set for further development.

Table 6.1 summarizes the link between the four phases of Intentional Development, the Essentials outlined in the previous chapter, and the associated Intentional Development Planning tools.

TABLE 6.1

Linking the Intentional Development Components to Intentional Development Essentials and Planning Tools

Intentional Development Component	Intentional Development Essential Associated with the Component	Intentional Development Plan Tool
Frame It	• Have a Planned and Targeted Impact	• Personal Business Case • Impact Map
See It	• Select the Key Competencies that will Meet Organizational Demand • Focus on the Critical Few Competencies • Create A Feedback-Rich Environment • Understand There's More Than One Path To Development • Make the Development Sticky	• Development Target • Learning Map
Own It	• Build It In; Don't Bolt It On • Reject "One Size Fits All"—Mass Personalization, • Create a Cadence of Development • Create a Feedback-Rich Environment • Match the Development Strategy to the Talent, • Never Learn Alone	• Intentional Development Action Plan • Intentional Development Tracker
Connect It	• Make the Development "Sticky" • Never Learn Alone • Create a Cadence of Development • Create a Feedback-Rich Environment,	• Intentional Development Evaluation • Career Development Plan

C. FRAME IT: ESTABLISH AND CONNECT THE BUSINESS CASE

> What matters is "Leadership for what purpose?" Leadership means getting the right things done.
>
> —Peter Drucker

When senior leaders ask us why it is important to establish a clear business case for development, we like to tell the story of what happens when Lewis Carrol's *Alice in Wonderland* comes to a fork in the road and asks the Cheshire cat in a tree,

"Which road should I take?"
The Cat responds, "Where do you want to go?"

Alice answers, "I don't know."

To which the cat responds, "Then it doesn't matter."

For development to be intentional, the road you take is critical—what we've described as having a "Planned and Targeted Impact" for development. In the previous cited research from McKinsey on why leadership development fails, "overlooking context" was a significant cause of failure. At the onset of any development effort, organizations must clearly answer the questions, "Why exactly are we doing this?" and "Why is this important for the business?"

Frame It is the stage of Intentional Development where we bring Mooney and Brinkerhoff's (2008) Impact Map to bear to establish a clear context for the development—a planned and targeted impact. As you will remember from our SDLD journey in Chapter 1, an Impact Map creates a clear line of sight that connects development on key competencies to critical job behaviors to important team and business outcomes. The basic outline for an Impact Map is shown in Figure 6.2.

In a coaching situation, we work jointly with the learner to discuss and confirm the Business Goal, Team Results, and Key Leader Results. (Key Development Targets are established in the next phase, See It.) In a cohort setting, we work interactively with the cohort members to build their own Maps (and to build accountability by reviewing their Maps with other members of the cohort). For team development, we facilitate a discussion with all the team members to generate, analyze, and agree on a Team Impact Map.

Impact Maps can be generated for an individual learner, for a team, or for a development cohort. Here are some examples (Table 6.2) of actual Impact Maps that we have developed with learners during the Frame It stage of Intentional Development.

Key Competencies to Develop or Enhance	Critical Leadership Challenges or Results	Key Team Results	Business Goal or Imperative
Specific competencies the leader must try out, use more of, use differently or use better to improve his/her performance	Key situations where different performance from the leader would lead to better team results	What the leader's team needs to achieve to drive the business goals	Critical few goals or imperatives needed to achieve the organization's strategy

←──

How does this happen?   **Why is this important?**

FIGURE 6.2
Impact map outline.

TABLE 6.2

Impact Map Outlines

D. Key Competencies to Develop or Enhance	C. Key Leader Results	B. Key Team Results	A. Key Business Imperative
• Building partnerships and working collaboratively across the business • Planning and prioritizing my team's work • Delegating effectively and removing barriers for my team • Developing and delivering clear and concise communications *(Identified in See It)*	• Make sure that the team hits our stage-gate targets • Assure that we are consistent and aligned on methods and messages • Assure that we cover all of our client opportunities in an optimal way	• Crafting a commercialization plan that locks in our clients • Streamlining the commercialization of our products	• Assure that we are the strategic partner of choice for our clients
How does this happen?			Why is this important?

D. Key Competencies to Develop or Enhance	C. Key Leader Results	B. Key Team Results	A. Business Imperative
Creating opportunities to cultivate innovation and drive engagement with my team Being more flexible and adaptable when addressing new business opportunities *(Identified in See It)*	• Create more time to focus on potential acquisitions • Empower the team to make changes in order for them to fully take ownership of projects, processes, and cross functional support objectives • Be seen as more willing to take on risk in situations where you are dealing with new opportunities	• Providing financial and organizational support on key initiatives dealing with new partnerships or acquisitions • Providing financial and HR shared support for new product launches and new customer acquisition • Implementing a new ERP system integrated across locations	• Exceed EBITDA goal of $15m • +14% Sales Growth through acquisitions and organic growth
How does this happen?			Why is this important?

D. Key Competencies to Develop or Enhance	C. Key Leader Results	B. Key Team Results	A. Business Imperative
• Relating effectively with a wide range of people in a variety of setting • More clearly communicating a vision of the City's future so that my team gets excited about it • Improving my ability to identify and assess new talent • Being more effective at holding others more accountable to their commitments *(Identified in See It)*	• Develop and communicate a vision for the management of the City for the City Council to adopt • Maintain and upgrade the knowledge base of the City workforce; • Help the City Leadership Team stay on track; • Communicate vision, plans and projects to the Council/ residents • Coordinate relationships with support staff and other City functions and related organizations	• Responding quickly and effectively to residents' requests for service • Working together to achieve the City's Goals • Providing input to projects and plans • Being the leaders for their teams with City goals forefront • Maintaining and improving their teams' skills sets • Embracing and implementing new technology and ideas • Working with all those we encounter using the core values of honesty, integrity, respect, teamwork, positive attitude, accountability, and stewardship	• Protect the property values, sense of place and way of life of the City • Maintain and upgrade the City infrastructure • Provide effective and efficient service to the residents of the City at the best value • Provide a great place for employees to work with a professional staff • Provide for efficient management of financial assets into the future
How does this happen?			Why is this important?

We find that the organizations that have the greatest success continue to emphasize the overriding strategic objectives of the organization and emphasize the requirements to achieve those objectives. Today's knowledge workers work within a context of what, why, and how—and they expect to be kept in the loop of the larger business context. Because of the clear context created by Impact Maps, the discussions that occur with learners around their creation and application are excellent "engageable moments" (an opportunity to engage employees, motivate them, and provide direction; Watson Wyatt, 2009).

LINE OF SIGHT

"Getting the business stakes out on the table and getting all levels of management aligned on the Line of Sight is critical for starting off in the right direction and gaining senior management's active commitment. Courageous trainers are skilled and relentless at clearly creating and communicating this business case."
(Mooney & Brinkerhoff, 2008, p. 41)

The Personal Business Case

The Impact Map clearly describes the business case from the organization's perspective. During the Frame It phase, we also like to work with learners to describe what we call their "Personal Business Case." Learners may have other interests or aspirations that could inform the context of their development efforts. We like to ask the learners several core questions:

1. Why is it important for you to work on development?
2. How will what you accomplish benefit you? What's in it for you to improve your knowledge, skill, or ability?
3. How will this help your career, and what do you want from your career?
4. How will you judge the success of this effort?

Once learners develop a draft of their Impact Map, we encourage them to share it with others who are important to their development journey, especially their direct manager. Peers, cohort members, mentors, and other "20s" may also be included. Sharing the Impact Map serves two purposes. First, this step serves as an important edit; others may suggest updates or additions to components of the Map that will make it more robust. Second, it builds accountability into the development process. Once a learner has reviewed the Map with others, there's no turning back! We find that Impact Maps also tend to be living documents in that they are often updated and added to throughout the development experience.

D. SEE IT: IDENTIFY ESSENTIAL SKILLS

In order to grow, we need to first have an understanding of who we are, what our strengths and capabilities are, and where we obtain the greatest gain for our efforts in improving. Building self-awareness is an important outcome of the See It phase of Intentional Development.

In a study conducted by at Cornell University and commissioned by J. P. Flaum at Green Peak Partners (2009), 72 senior executives from across 31 companies were interviewed and surveyed. The results were mapped and, in follow-up interviews with their managers, additional information was obtained about these leaders' self-awareness. The results indicated that a high self-awareness score was the strongest predictor of overall success. Leaders who understood their strengths *and* understood where they could improve were likely to grow in their role. Furthermore, they recognized how to complement their own capabilities by hiring around their weaknesses by finding people who could fill the gaps in areas where they were not strong. This, of course, is a dynamic rather than static process, in which the individual is always looking to get better. See It is more than just knowing oneself. It is really about understanding yourself, identifying where and how to improve yourself, and being open to grow and learn—the Growth Mindset we referred to earlier. See It is not all about being able to just identify your strengths but also to look at weaknesses. We don't necessarily endorse the use of the word "weaknesses," not because we want to be politically correct, but because we don't think it is a useful way of thinking about our skills.

The truth is that for most of us, we are competent at most of the activities or competencies needed at our jobs. Our natural tendencies drive us to do what we do best, and we use mostly our strengths in that effort. In addition, calling these areas where we don't excel "weaknesses" contributes to building a negative mindset, which usually leaves people overly focused on "what I don't do well" as opposed to "what I do well." We prefer to think of them as "opportunity areas." Thinking about development needs in this broader sense also sets the stage for considering a much wider and varied range of improvement strategies such as the following (a preview of some strategies we'll discuss in Own It):

- Take a skill that you are already good at accomplishing and improving the skills to be masterful.
- Work on something that you do okay and bring it to a state of improved competency.
- Dial down a strength that you might tend to overuse and on which you've become over reliant.
- Save energy, time, and results by not contributing in an area where you know you are not strong.

- Work on a skill that you've never had to use before but is now critical to success in a new role.
- Explore a skill that you think you do really well, but others don't think you do that well (called a blind spot.)
- Get more feedback on a skill that you don't think you do well, but that others think you are good at demonstrating, called a hidden strength.

While self-awareness is a critical factor in our professional growth, insight alone is not enough to grow as a professional. We must get regular feedback from others and even more importantly, we must be able to accept that feedback and that in and of itself is a critical (and essential) skill to develop. Consider what happens when you know your annual performance review is coming up. It's pretty typical for people to gird themselves for that event. As we described earlier, the threat response is activated and we are ready for action so we prepare our explanations for when the discussion turns to missed deadlines, flawed project outcomes, or revenue targets misses. Clarifying questions are asked about whether uncompleted projects were truly priorities and justifications are lined up to ensure that the other perspective is well represented.

While it may be the nature of the performance review process that sets the stage for this kind of reaction, it is no less under the control of the employee to consider ways to take that information they are receiving in a more helpful and supportive manner. The manager, of course, can ease the stress of

THE ESSENTIAL SKILLS

It's time to retire the idea of "soft skills" This term that has been used for the past 30 years or so is always said as a counterpoint to the notion of "hard skills," which is typically related to financial or other objective data. Over the years the use of the term "soft skills" is usually said in a disparaging way as in, "we recognize that these skills are probably important, but they are *not nearly as important* as the hard skills of being able to do a financial spreadsheet or create a process flow chart." What we now know from research from Harvard, the Carnegie Foundation, and Stanford University among others is that what are referred to as "soft skills" would better be named the "essential skills" for success at work. These skills (such as communicating effectively, demonstrating influence, and managing conflict, among others) define success for over 85% or all jobs while the technical or "hard skills" only define success for about 15% of all jobs.

that situation by making the feedback developmental in nature and geared toward supporting and improving success for the employee except in cases where performance truly is an issue. In return, employees want to work to see the feedback as useful and not as criticism about what is wrong with them. Shifting one's mindset to recognizing the value of effective feedback puts that information into proper perspective and allows it to be used for more understanding and ways to become an improved performer at work.

In the See It phase, two critical pieces of information merge from the Strategy-Driven Leadership Development Model (SDLD). First are the mission-critical leadership competencies which will help the company succeed (the importance of the competencies). Second are the competencies the individual employee possesses or needs to develop in order to meet the demands and expectations of the firm (the level of skill or competence).

As we described, the importance of different competencies is determined in our approach in the Organization Demand modeling process, where strategy is translated into leadership competencies. If you are starting at a different point in the model and the importance of competencies has not been clearly established, then that data will need to be gathered for the learners either through a separate assessment step or by asking raters to provide feedback on importance in a multi-rater feedback (360), assessment interviews, or other method.

We have also helped organizations to initially determine the relevant importance of different leadership competencies by working backwards from the talent review data. We do this by parsing out of the talent data the competencies that have been identified as strengths of the highest performing and highest potential employees. We encourage the organizations to use this approach as a starting point and to eventually do more formal work around translating strategy to competencies.

The level of competence can also come from a variety of sources. Of course, an important source of feedback could be from the data gathered in talent reviews. By "data" we do not mean telling the employee his/her specific Development Strategy (i.e., "You are an Emerging Leader") but rather sharing the behavioral observations that came from multiple observers and the calibration discussions. We find that not sharing or sharing the feedback from talent reviews often becomes a policy decision for organizations depending on previous practice *and* the amount of training and experience that their leaders have in providing effective feedback. Providing feedback from talent reviews can also be perceived as imposed rather than requested.

TO TELL OR NOT TO TELL

Do you tell High Potential employees that they are seen as High Potentials by the organization's executives? What do you tell people that have been assessed in a talent review?

These are questions that often come up in our work and have been a regular topic in talent management articles. From our standpoint, we believe that everyone that is included in a talent review should get feedback. However, the context and content of the feedback is critical

So that the feedback in placed in some context, we recommend that the feedback be accompanied by an Impact Map or at least the discussion be framed by the Impact Map outline of "Why is this important?" flowing to "How is this accomplished?" Putting the feedback in a broader personal and organizational context assures that the feedback perceived as relevant and not just imposed on the employee for no reason. The feedback should also be followed by working with the employee to build an Intentional Development Plan (where relevant) so that the feedback is seen as actionable and accountability is established for following up.

From a content perspective, we never suggest that the feedback includes just the Development Strategy category, such as High Professional or Emerging Leader. Rather, there are specific scripts that flow directly from the Development Strategy supported by strengths and development opportunities described in behavioral terms (competencies).

We have found that the richest form of developmental feedback comes from combining the talent review observations with properly designed and implemented multi-rater feedback (360s). As we have described, the talent review identifies who would benefit the most from Intentional Development. The multi-rater feedback adds another layer of feedback . . . and the opportunity for the learner to request the feedback. The initial development discussion with a High Potential, Adaptable Professional, or Emerging Leader would go something like this (very much paraphrased), "I and other leaders at ACME Inc. think you have a very bright future with the company. To that end, I would like to offer you the opportunity to participate in a focused development process that includes getting some additional feedback from your team, your peers, and other leaders. What do you think?" With the learner's buy-in, the multi-rater feedback can be gathered and then combined with talent review observations for a very rich picture of the individual learner's capability.

While we are on the topic, here are a few important guidelines to follow when employing multi-rater feedback.

- Avoid building 360s in-house unless you have the psychometric and technological expertise on your staff that understands what is required to create a valid and reliable measurement tool. There are several well-designed tools available that have extensive norms and well-tested delivery methods.
- If you opt to use 360 interviews instead of a survey, the interviewers must be trained in behavior-based assessment and have an in-depth knowledge of your competencies.
- The individual learners should ask those giving them feedback by asking them directly, by phone or face-to-face. We like to provide guidance to the learners on how to approach their raters and maybe even a script to follow.
- All raters should go through an orientation so they clearly understand the purpose of the feedback, how the information will be used, and what constitutes effective feedback. We have found that ignoring this step reduces the response rate and the amount and quality of written comments (if requested).
- All raters should be guaranteed anonymity, except the learner's boss.
- Those that review the feedback with the learners should be skilled at interpreting and feeding back the results. If the person giving the feedback and providing coaching support to the learner is the same, even better.

Development Targets—Key Competencies to Develop or Enhance

This is an important step in the See It model in that it provides the frame from which an employee can select where he or she wants to put development efforts. Several factors play into considering which competencies to select.

- Which competency, if developed or enhanced, will yield the greatest benefits for the learner and the organization (referring to the Impact Map)?
- How important the competency is for success in their current or near future role?
- What feedback from 360s, talent reviews, and/or performance reviews has been highlighted?
- Which competencies are differentiating for the learner compared to other people in his or her department or current team?

- What current work challenges would be a good test bed for the competency; is there an opportunity?
- How difficult will the competency be to develop; will the competency require multiple development experiences?
- Can the learner "see it"; can he or she envision what success and the impact of success would look like?
- Is the learner excited/engaged by focusing on the particular competency?

In the end, the learner has to have a clear reason or meaningful logic for selecting the developmental target, or it's not beneficial to move forward. As we like to say, the development logic "needs to tell a good story."

We use a specific format to help the employee see that the competencies to focus on are development targets for them to set their sights upon. The structure might look like Table 6.3.

TABLE 6.3

See It: Development Planning Targets

Development Targets	Development Reason or Logic
Name the competencies that you want to try out, use more of, use less of, use differently, or do better	How will these efforts make a difference to your firm and for your professional development?
Competency 1:	
Competency 2:	
Competency 3:	

And a completed Development Target profile might look like that in Table 6.4.

Intentional Development Plan: The Learning Map

After learners have targeted specific competencies and their logic for selecting them, we ask them to build a Learning Map. The Learning Map is the epitome of "intention." It helps the learners identify their current skill levels on the competency, where they would like to see their skill levels to be, and the payoff from achieving a new skill level (capitalizing on three of

TABLE 6.4

Development Planning Target Example

Development Targets	Development Reason or Logic
Name the competencies that you want to try out, use more of, use less of, use differently or do better	How will these efforts make a difference to your firm and for your professional development?
Competency 1: Promoting Collaboration among Team Members	Our team does not work very effectively and there are often tasks that are done by multiple people at the same time. By working more collaboratively with each other we will save time and decrease repeat actions that rob us of valuable work time.
Competency 2: Being Resilient	Our team operates under a great deal of stress and pressure and we've burnt people out who've moved on to other organizations or teams. We could improve our retention and team focus if we could learn to handle difficult situations more effectively
Competency 3: Business Acumen	I operate on the customer side of our business and do not usually understand the financial elements of our business. I believe I can be more effective in my role if I understood the financial and sales drivers that impact our ability for our firm's success

Knowles's keys to adult learning). This is also the stage in the process where the use of Intentional Development Mantras kicks in. Table 6.5 provides an outline of a typical Learning Map.

Of the different parts of the Learning Map, the Intentional Behavior Targets are probably the most important, followed by the Intentional Development Mantra. The Behavior Targets set the learner's intention and prepares his/her brain to learn something new. It also serves the important purpose of answering the questions, "When have I made sufficient progress on this competency?" or "When is it time for me to create a new Intentional Development Plan?"

As we've outlined before, the mantra provides a simple rule of thumb that can trigger a new behavior and make the development stick. Learners use the mantra to keep their eye out for development opportunities, to remind them about what new behavior they want to practice or try out, and to see if the new behavior had the desired effect.

TABLE 6.5

Learning Map Outline

Why this Competency was chosen as a Development Need	The competency that is the focus of the Intentional Development Plan
	Description on the development need; why the target competency was chosen above all others.
Benefit from Improving his/her Skill	Description of the positive impacts as a result of becoming more skilled at the competency. What's in it for the learner to take on this challenge? How will s/he *feel* when s/he is more talented at this competency?
Current Competency Description	Description of the learner's current skill level on this competency in terms of behaviors s/he exhibits now. Consider boss feedback, multi-rater feedback, and/or the output from a talent review
Intentional Behavior Targets	Describe the skill level on the competency that the learner would intend to achieve. Describe the competency at a more talented or effective level. How would you describe someone who is exceptional at this competency? What would it look like?
Intentional Development Mantra	A statement(s) in the form of "If I am in situation X, then I will do Y in order to achieve goal Z" that the learner can use as a reminder/trigger for applying or trying the target competency.

Tables 6.6 and 6.7 give a couple of examples of completed Learning Maps.

Getting people to think in terms of business competencies and to be able to name those competencies is a huge leap in leadership and employee development as it moves us away from the negatives of how people feel they are evaluated and toward a more neutral stance that relates to the skills associated with success in the workplace. We may never be able to get away from the personality descriptions that people use to talk about their employees but, by bringing a competency model to the heart of the discussion, both the employee and manager begin to get on some even footing. Consider the discussion that a manager might have with a relatively new hire who is not strong on understanding how the business operates and perhaps is not familiar with manufacturing processes. He was hired as a relationship specialist for this firm's customers.

"John, I heard from Tamara over at Lighten Chemicals, and she told me that when she was talking to you about the need for speedier deliveries so

TABLE 6.6

Learning Map Example I

Name	Pat Example
Date Created	01/15/2019
Date to be Completed	03/15/2019
Competency Name	Problem Solving
Why this Competency was chosen as a Development Need	This skill is critical to our product development stage gate process. I need to become a more disciplined problem solver so that we have more success with projects.
Benefit from Becoming Skilled or Talented	Our team hits our stage-gate targets and our product launches are more successful. I contribute more to our product development and help the team hit stage-gate targets more effectively. I am seen as a facilitator of the results and not a barrier to our performance. My team shows more respect for me and exhibits less frustration. My boss doesn't give me negative feedback on problem solving anymore.
Current Competency Description	I am okay at solving problems but am not disciplined about it. I tend to jump to conclusions to quickly and don't consider enough information or evidence. I am new to managing projects by using the stage-gate process. I have to go back and rework problems because my first thoughts weren't effective. I've been told that I tend to oversimplify problems and go with my gut instead of good analysis
Intentional Behavior Target	I look beyond the initial evidence/facts and just not stop at the obvious. I use a standard step-by-step approach to problem solving that is less based on intuition. I spend less time on rework.

Your Intentional Development Mantra

Situation	What You Will Do	So that . . .
When I'm working on solving a product problem,	I'll slow down and follow a step-by-step process	so that I have less rework and my team has more successes

that they could ensure that their throughput could be increased, you asked her what 'throughput' was, and she was shocked. She told me she had to describe to you their manufacturing process and that you asked her for a tour of the facility so that you could become more familiar with their operation. You've had this account for six months! How could you not understand what their business is about?"

TABLE 6.7

Learning Map Example II

Name	Taylor Learner
Date Created	10/1/2018
Date to be Completed	3/15/2019
Competency	Making quicker and better quality decisions
Why this Competency was chosen as a Development Need	From my 360 feedback, this was one of the lowest rated skills for me but it is also one of the highest in importance for my role. My boss has also mentioned this need to me in the past.
Benefit from Becoming More Skilled	In my current role, it is critical that I am able to make good and timely decisions that keep the organization moving forward. I work in our largest and fastest growing market, so the decisions I and my team make are very critical to our overall success. A comment from one of my bosses was that if I make stronger recommendations, I can make the decision process simpler for those up the chain and, by doing so, I will help make everyone more efficient. I will also become more respected by others for displaying better judgment.
Current Competency Description	Approaches decisions haphazardly. Makes decisions based on incomplete data or inaccurate assumptions Ignores different points of view Makes decisions that impact short-term results at the expense of longer-term goals.
Intentional Behavior Target	Makes high-quality decisions, even when based on incomplete information or in the face of uncertainty Spends less time analyzing and more time getting others' input Actively seeks input from others to make more timely and well-informed decisions. Is more effective at separating opinions from facts

Your Intentional Development Mantras

Situation	What You Will Do	So that . . .
When I need to make a decision on buying a new location that impacts our region	I will spend less time doing analysis on my computer, more time talking to our partners and vendors and work through a step-by-step process with my team	I make decisions less haphazardly and make recommendations that make it easier for my boss to make the final decision

While a competency-based discussion may not be any less intense, it may provide a different route for addressing the issue that is of concern for all parties.

"John, I heard from Tamara over at Lighten Chemicals, and she told me that when she was talking to you about the need for speedier deliveries so that they could ensure that their throughput could be increased, you asked her what 'throughput' was and she was shocked. I did not realize that you were lacking in some basic understanding of manufacturing processes. Perhaps, you could tell me in your own words what you understand their business to be about. I think there are some things we can do to make sure you are more knowledgeable about business terms and practices."

"Business acumen" would be the competency lacking in John's behavioral skill set. Having knowledge of business practices and the associated words is a learnable skill. In fact, by identifying this as an issue for John, his boss now has the opportunity to improve an area of John's skill set that will benefit his work with all his customers. We would see improvement in John's business acumen when he demonstrates:

- Comfort in using business terms and using them appropriately
- Ability to speak to how their actions impact business results
- Ability to represent the organization's product or service to outsiders
- Ability to help his customers understand how his products and services help their business
- An understanding of strategy and the ability to apply it in a tactical manner

You'll also notice that See It includes asking the important question of why developing this competency will "make a difference to your firm and for your professional development." Expanding the perspective of See It to include how this effort will also help the firm and not just the individual expands the perspective of the employee to see that development activities benefit that person and the organization.

We are ready to move on to maybe the most challenging aspect of the Intentional Development process, which is Own It.

E. OWN IT: BUILD THE SKILLS

By three methods we may learn wisdom: First, by reflection, which is noblest; Second, by imitation, which is easiest; and third, by experience, which is the bitterest.

—Confucius

We've outlined the importance of creating a targeted business and personal impact for development (Frame It) and we've described more detail on identifying the particular competencies essential for development and a concise map for the learning voyage (See It). Now we come to Own It, the step in Intentional Development where most of the actual developmental action takes place. The focus is on plotting out and implementing an Intentional Development Action Plan and then locking in the learning through Intentional Development tracking.

Own It is built directly from the 70–20–10 learning model and engages several of the core essentials of Intentional Development, namely:

- Build It In; Don't Bolt It On
- Understand There's More Than One Path to Development
- Create a Cadence of Development
- Create a Feedback-Rich Environment

Intentional Development Plan: Development Action Plan

Understanding that development occurs following the 70–20–10 model is one thing. Using it in regular practice is something else and can be challenging to accomplish. When left to their own devices, learners and their managers tend to default to "attend such-and-such training class" as their development plan. To make 70–20–10 come alive, we use an Intentional Development Action Plan that starts with identifying a specific task(s), assignment(s), project(s) or other on-the-job learning opportunity (the 70s). We then link other people (the 20) and just-in-time learning (the 10s) to the developmental experience.

The format of an Intentional Development Plan follows the 70–20–10 outline, identifying development tactics for each component of the model. A basic outline of a Development Action Plan is described in Table 6.8.

In some cases, it may take more than one 70 if the target competency is not easily acquired. Most jobs have plenty of challenging opportunities where the learner can try out a new behavior or use more of the behavior, use it differently or use it better. However, we sometimes find that the learner's current role is not very developmental (i.e., they've done the job for a long time or have worked similar roles previously). In those cases, we have to work with the learner (and their manager) to change the job or find a challenging assignment that requires the use of the target competency.

Once learners have created their Impact Map, their Learning Map, and their Development Action Plan, we encourage them to review the Intentional Development Plan with their manager (and/or with their

TABLE 6.8

Development Action Plan Outline

Development Tactic	Development Action
Assignments, Tasks, Challenges, or On-the-Job Learning (The 70%)	In what specific situations could the learner try the competency out, use this competency more, use it differently or use it better? Where can the learner practice or experiment to develop the level of skill on their target competency? What experiences will help build the target competency? What jobs, day-to-day tasks, assignments, projects, challenges, and/or practices will help develop the skill? Refer to the Key or Critical Leadership Challenges on the Impact Map for suggestions? When will the action be accomplished and what results are expected from it?
Coaching, Mentoring, Role Model or Accountability Partner (The 20%)	Who is particularly good at/expert at the development target competency? Who would be a good coach or mentor for the learner? Who would be a great source for additional feedback? Other people that could be a resource for the learner as they take on the assignment or challenge (the 70%). Who is already skilled at the competency? Who would be a good mentor and/or accountability partner? Who handles the Critical Leadership Challenges (see the Impact Map) well? Who would be a great source of additional feedback?
Self-study or Coursework (The 10%)	What does the learner need to learn more about or use to keep focused and be successful in the Development Action Plan? Where to learn more about the specific competency? What resources can be used to learn more? Don't forget podcasts, YouTube videos, or Ted Talks.

Development Cohort and Accountability Partner). The review serves two purposes—to get additional ideas and suggestions to enhance the Plan and to build a commitment by the learner to follow through. The learners also need to connect with any 20s they've outlined in their Action Plan to ask for them to play the role outlined for them in the Development Action Plan.

The next step in the process is the "just do it" phase, where the learners put their Development Plan into action. They complete the self-study 10s. They watch for situations pinpointed in their mantras where they can try out a new behavior. And they get additional feedback and support from their 20s.

Most agile learners tend to jump right into their Development Plans. Other learners might need additional support or encouragement. This is where the role of a coach or manager can be helpful. It also helps to make sure that target dates are included in the Development Action Plans.

As we described, Intentional Development work is not a one-and-done but requires a regular cadence of development—try something, reflect on how it plays out, look for the expected impact, adjust if necessary, and try again. We encourage learners to work on their plans on a daily or weekly basis. These kinds of micro-learning opportunities create a consistent learning environment and are critical to building a habit for change rather than hoping that willpower will save the day. One executive client set aside time on his calendar every Friday, what he called "5 At 5 On 5" which meant "5 minutes at 5:00 on Friday," where he pulled out his Intentional Development Plan, reflected on the progress he made that week, and plotted out what he was going to try out or focus on in the coming week. He made excellent progress on his development and quickly locked-in-the-learning, which we believe can take weeks and not months and certainly not years.

While experience is critical to learning, research has shown that so is regular and deliberate reflection on the learning experiences we are navigating as well as documenting insights, which usually means writing them down somewhere whether it be in a journal or in your calendar. In carefully designed field study and laboratory experiments, Di Stefano, Gino, Pisano, and Staats (2016) found that spending just 15 minutes a day reflecting on and writing about a learning challenge resulted in a significant improvement in performance by call center workers and increased customer satisfaction. An interesting corollary from the findings of the same study was that, at some point, the benefits of accumulating additional experiences were less than taking the time to reflect on experiences that the learners already had. And, when given a choice, learners tend to opt for having more experiences versus spending time in reflection, even though that is not optimal for learning. Reflection makes a difference.

As coaches, we will often send out regular Intentional Development reminders to learners to encourage them to reflect on their developmental experiences. The reminders encourage the learners to track their progress and also include Intentional Development Tips and Techniques. We have tools that we use to assist learners in creating a cadence of development and to make sure that they lock-in-the-learning from their development experiences. We call it the Intentional Development Tracker.

Intentional Development Tracker: Lock-in-the-Learning

We've seen various tools to help learners track their progress. Keeping a development journal is one approach that has been effective. We've enhanced the journaling process by making it more intentional. We ask learners to summarize each development action they take using the format outlined in Table 6.9, which we call the Intentional Development Tracker (also diagramed in Figure 6.3).

Owning Our Successes

We were attending a conference several years ago and the leader was discussing how we all could improve our self-esteem by acknowledging our strengths and successes. He commented that most of us are raised to be humble, which is a noble quality, but sometimes it goes a little too far to the point that we do not even acknowledge all the things we do well.

TABLE 6.9

Own It: Intentional Development Tracker Outline

Context	The development target described in terms of a competency or behavior and the development mantra from the Learning Map
Experience	A description of the learning experience (from the Development Action Plan) that was just attempted. What was the specific new or different behavior that was attempted and in what situation was it tried out?
Reflection	Summarize what was learned from the developmental experience in the form of "when I do this, this happens or didn't happen" and why.
Impact	A description of the impact of what was learned (on the learner, the team, or the company). Is it what was outlined in the Impact Map or was there a different outcome? How did the experience impact the learner?
Intention	What the learner would do differently next time based on what s/he learned from this experience. Outline how to apply the lessons of this experience in the future, in what situations. What new mantras or intentions may be suggested from this experience?
Planning	Based on this development experience, are there any changes or updates that need to be made to the Intentional Development Plan (Impact Map, Learning Map and/or Development Action Plan)? Has the Intentional Behavior Target described in the Learning Map been achieved? If not, what events, situations or experiences in the coming week(s) could be used as an opportunity to try to do something differently, more of, less of or better in line with the Development Plan? What action could be planned to take next? Who could help with the opportunity (provide support or suggestions, observe and give feedback)?

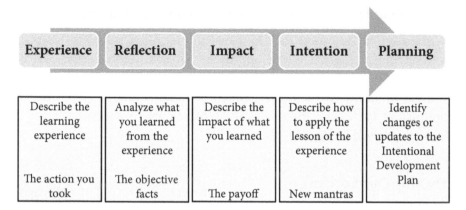

Experience	Reflection	Impact	Intention	Planning
Describe the learning experience	Analyze what you learned from the experience	Describe the impact of what you learned	Describe how to apply the lesson of the experience	Identify changes or updates to the Intentional Development Plan
The action you took	The objective facts	The payoff	New mantras	

FIGURE 6.3
Intentional development tracker.

By example, the leader asked a woman in the audience about any special talents, skills, or experiences she possessed. She was a successful businesswoman, but when was asked to detail a specific capability, she couldn't identify anything. He went on to ask her about her college experiences and she shrugged and mentioned that she had attended West Point. "Oh," he asked, "when were you at West Point?" She told him that she went there is 1976. He commented that there probably were not many women attending West Point back then and she replied that she was in the first class of 119 women to attend West Point and that after that she had gone on to a successful career in the military retiring with a rank of colonel. "Oh" he added, "I guess you really didn't have very many unusual talents or skills to share with us," to which everyone in the audience including the retired colonel began laughing out loud.

Claiming our competencies, strengths, and successes can be one of the most challenging aspects of this or any development model. It is easier for us to focus on our weaknesses and look for how we can improve than to see our capabilities and build from there. In part, it is because our brains are wired to protect downside risk that we might face in challenging situations. Consider early humans walking the savanna in search of the woolly mammoth. They're out hunting but always looking behind their shoulder to make sure that the saber-toothed tigers are not right behind them looking for their next meal. This protective mechanism, one we still possess today, is what we have referred to previously as the negativity bias.

The Challenge of Acknowledging Our Weaknesses

While it might be easier to acknowledge negatives rather than positives, this has its limits as well. It's almost as if our humility has a limit to how far it will go. If we get criticized or feel as if we are being unfairly treated in terms of where we need to improve, we'll question that issue as well. Once again, we can turn to brain research and see that there is a cognitive bias that explains this behavior as well and it is known as the "self-serving bias." The self-serving bias is used to make sure that our egos are not hurt too much by negative feedback. The self-serving bias can also be used by those who may not want to take full responsibility for their actions or who may attribute failures to others before they take it on themselves.

One young high potential employee we knew had just received her performance review from her boss. She was dismayed to find out that her boss told her that she thought she was too negative about things and often impeded progress that the team was trying to make by consistently insinuating that there would be a problem with some aspect of the project. Her manager gave her several specific examples but when we got together to discuss the matter, she told us that she always felt that it was her responsibility as a leader to point out flaws and errors with the team. She had no idea that her efforts to strengthen her team's actions were being viewed as negative by her boss. She begrudgingly accepted the feedback but persisted in believing that her actions were in the best interest of the team. Unfortunately, instead of trying to change her perspective and actions in how she approached meetings, she instead became more silent and did not contribute to the conversation even when she had valid points to register. Her inability to break the bias that had served her well in the past eventually led to her being let go by the firm.

Owning It All and Changing It All

Good or bad, strong, or weak, we need to own it all. Overcoming these cognitive biases is not easy, but changing the way we think is doable and when we do, we create new biases that focus on things like positivity and self-responsibility. These are important for development as they increase our self-awareness and they also create an important shift in how we approach our work.

Brain research has identified three steps that we can take to change our biases and mindset so that we can think and behave in a different way.

These include (1) becoming aware of our biases and how that affects our mindset, (2) raising this thinking to a level in our own mind to where we are concerned about it, and (3) replacing the biased thinking with mental and behavioral responses that more appropriately match our actions to our stated values. For leadership development that translates to recognizing the leader's role in creating the greatest opportunity for growth by, of course, being intentional about his/her actions and the actions of his/her team. Strategic leadership development is a team exercise in which everyone plays an important role. By engaging all parties, the opportunity for success is much greater, but all must know their roles and be intentional about them. As one of our clients says, "Development is employee owned, manager enabled, and company supported." So, what are some of the specific steps we can take to change these mindsets?

EMAIL MADNESS

One of our clients had over 5,000 emails in her Outlook and after hearing all the rationale for why she couldn't delete some of them (they were several years old), we examined some of them from just the last month, and she realized that the critical nature of most of them had long since passed. With some trepidation, she went ahead and deleted the vast majority, just holding on to the last few months' to make sure she wasn't missing anything.

We helped her establish a new system for reviewing emails and she was breathing a big sigh of relief (and getting more work done as she was not ruminating about email!).

Habit vs Experimentation

At one level, we can think of cognitive biases as habits, behaviors that we engage in on an unconscious level. Habits like brushing our teeth and how we check our email are very helpful because they are purposeful, efficient, and effective. They need little thought and get the job done. The downside of habits is that they may keep us in a box that is not the most efficient or effective way to approach the task. For example, above we mention that our email checking behavior is habituated. Most of us have our routines for checking email. It may happen whenever we hear the email ping or we check every five minutes just to make sure that our boss has not sent us a note with a high priority flag attached. Regardless, for most of us, the research would suggest that we are not efficient about email in that we are far too often distracted by the sound or sight of incoming mail that we feel obligated to check and which usually distracts us from the task at hand. Changing that habit may be challenging but is definitely doable.

Brain research suggests that by experimenting with some other approaches, we create new neural pathways that we may find create more effective behaviors (Swart, Chisholm, & Brown, 2015). When this topic comes up in our consulting work (and it often does), we provide our clients with four to five different email models to try out, including some that they have probably never considered such as just checking your email three times per day, and to see how these might work for them. Of course, beyond the experimentation of trying out a new approach, our goal is to make them uncomfortable so that they get out of their typical routine and explore a new idea that might rewire an approach that wasn't working all that well for them. All it takes is a willingness to try something different to see if it works better. Of course, if our approach did not work better for them, then they are free to return to their former way, but at least we got them to try something different and begin the exploration of making changes and thinking differently.

We will also use this model to change our leaders' approach toward owning their strengths and weaknesses. Their See It exercise gives them that information, but that is not enough. In our cohort groups, we get leaders to share their strengths and weaknesses with others and use that information as feedback for themselves and from others. We might, for example, put them through an exercise such as the "Strengths Shield" where they name their five top five strengths, drawn from the results from the Clifton Strengths Finders Assessment and then draw a picture of each of their strengths on the shield. This exercise might be followed by an "Opportunity Shield" (See Image 6.1) where they would do the same activity but instead would focus on their weaknesses. Each member would then go around the room and introduce him/herself to their colleagues using his/her Strengths and Opportunities Shields to discuss his/her leadership skills. Getting people to announce their strengths and opportunities for improvement in front of their colleagues helps people rapidly own their behaviors.

Experimenting does require a certain amount of willingness on the part of the leader, but trying something new means that we don't have guarantees about how things are going to work out. We suggest that we make the

IMAGE 6.1
Opportunity shield.

exercise fun by seeing it as an adventure with minimal downside risk A few examples might include:

- Talking about failures and having everyone share a failure experience and how it turned out over time
- Using improv or an improv game to get people out of their usual way of interacting and laughing a bit about their work
- Discussing how their leadership affected one of their direct reports and how that contributed to success within the organization
- Having a leader acknowledge success by his/her team members, naming specific behaviors that were taken and what specific steps his/her team members took to create the win

Seeing Success in Others

Building the muscle of acknowledging success comes with practice. Our natural tendency to diminish our own capabilities is a part of our nature to not boast or be self-promotional but it does limit how we see ourselves and how others see us in the workplace.

We were working with a highly successful leader recently who self-admitted that she did not think of herself as being particularly excellent in her work despite feedback from her boss and colleagues. She was the go-to problem solver for her organization, and whenever an issue arose with a vendor, customer, or consultant, she was called upon to solve the problem. She had trained her team to work like first responders who jumped on a problem before others barely knew what was going on. In addition, during the past year she had been given a mission-critical project and had empowered her team to move the plan ahead quickly. The program was successfully launched well before anyone thought s/he could get it done. As we reviewed her successes during the past year, she could not admit that she had been successful beyond anyone's expectations. She kept saying that she had worked hard and had been really lucky to get as much done as she had in the past year.

We wanted her to acknowledge her own contribution to the success during the past year and asked about how well her team had done in achieving these goals. With that opening, she was ebullient about describing all the hard work and success they achieved during the past year. As our discussion ensued, we were able to have her share the outstanding efforts made by each of her team members and it became inevitable that by the end of the conversation, she had to acknowledge her own role in her team's success. In fact, not only had she recognized her role as a leader in empowering the team but in her own role as the strategic leader. This brought us nicely back to her own development work, which was to move from being a tactical leader to one who drives strategy.

Seeing success in others is always an easier task than it is to see it in ourselves. We are ready to acknowledge other's good deeds knowing that it reinforces the effective behaviors that lead to wins. There are both scientific principles as well art to giving feedback, and we'll be discussing them further down the road in this book when we get to coaching, but for now a few pointers are important here as they also guide us to how we can acknowledge the success in others and, by corollary, ourselves.

See Small Victories

The notion behind leading and lagging indicators was first developed when the idea of the Balanced Scorecard was described. Lagging indicators are typically "output" oriented, easy to measure but hard to improve on because you may not always know what caused the result. Leading indicators, however, typically look at what are the inputs that lead to results and provide a means of creating success. An easy way to think of leading and lagging indicators relates to weight loss efforts. Stepping on the scale is a lagging indicator which often creates a great deal of frustration and aggravation for folks if the numbers are not going down. Of course, the leading indicators include decreasing calories and increasing exercise, and if people charted those activities as well as the tracked their weight, then loss would be an inevitable event.

Say for example, that one of your salespeople is really good with her customers when she gets in front of them but becomes easily discouraged if she can't get the meeting set up rapidly; the behavior you want to reinforce is persistence. You don't necessarily want to hear about every phone call effort that she's made to reach out to this prospective candidate, but you do want to review her plan about building persistence through (1) having regularly scheduled calls, (2) having a solid script to make sure she connects with the buyer, (3) keeping notes in your CRM software so that she can recall her last efforts, and (4) any other approach that will help keep her spirits up for one more try. Checking in with that staff member about building persistence through these small steps not only builds the muscle of persistence but helps you as the coach to recognize and to respond to the leading indicators that build that success.

As you as a leader begin to notice these small wins, the skill of recognition begins to translate into what you can see for yourself in terms of your leading and lagging indicators. Small wins like clearing out your email at the end of the day reinforces a sense of organization. Reiterating strategy at your team meeting builds alignment with your team. Repeating organizational messaging multiple times helps relieve ambiguity among your team members. Small steps that build toward the goals you want to achieve are accomplished through these efforts and as you see these wins, the shift in your own thinking about who you are as a leader begins to shift as well.

CHANGE THE WORLD: THE POWER OF HABIT

Navy Admiral William H. McRaven served as the ninth commander of the US Special Operations Command (Navy Seals) and is now the chancellor of the University of Texas. In his 2017 commencement address at UT, he spoke about one of the most important way for anyone to change the world and that was to start off your day by making your bed. He told the graduates that by starting off each day by building on that small habit, you will find that you can create success after success during the course of the day. That was one of the most important lessons he learned when training to be a Navy Seal. In an enlightened conclusion to this part of his speech, he pointed out that even if you had a miserable day, you would return to a beautifully made bed, which would give hope and inspiration that tomorrow would be better.

The power of habit.

Being Behavior Specific

Being smart is good but it is not sufficient for success. Perhaps more important than intelligence is making a concerted effort and understanding how the effort led to mastery of a task. In research done by Carol Dweck (2016), who developed the idea of "growth vs fixed mindsets," children who were told they were smart and should do well on intellectual tests not only did not perform well on examinations but also had all kinds of anxieties and worries about their performance. On the other hand, children who were praised because of their effort saw more success and, perhaps, more importantly, saw their success as being a part of their own ability to take charge of their actions. As a result, these children developed the skills needed to take successful action.

While that approach is clearly beneficial in the classroom, it is not used enough in the workplace, where effort must be accompanied by success. To make that leap, leaders must do more than be praiseful of their employees. Saying things like "I expect more from you. I hired you because you are smart," fails the first test of thinking success is tied to intelligence. "You may just have to put more effort in. I stay late to get things done and I work on weekends; maybe you should too," may create a greater sense of motivation (although it probably won't) but still misses out on what it is the individual should do to be more successful.

The real key to helping your employees recognize that the behaviors of success lies in identifying for them the specific actions they took that led to success. You may want to assume that your employees know how to set priorities, filter emails, manage a spreadsheet, deal with an angry employee, set expectations with their customers, and any other number of specific skills, but you might be wrong. Helping them see that when they sit down each week and prioritize their key activities that will drive business success and how their prioritization process establishes time and effort activities for them is what helps them to stay on task to get the work completed.

While everyone in the workplace today is operating on some kind of team, it is not correct to assume that everyone knows *how* to be on a team.

It's very possible that if your work colleagues were not in the marching band in high school or college or drama club, or participated in Girl or Boy Scouts, or played a team sport, they might not know how to be on a team. Some people were loners during these formative years, while others were perfecting the skills of shared responsibility. There is a high probability that if some of your employees were never on a team when they were younger, they may not know how to participate on a team today and may need to learn the skills of team effectiveness; yet we often assume that everyone knows how to be on a team.

F. CONNECT IT: LEVERAGE NEW SKILLS

In the Intentional Development Process for a learner, the Connect It stage concentrates on assuring that newly skills are locked-in, that key developmental experiences are positive (additive) to a career, and that the stage is set for further development. We also take the opportunity to reflect on the Intentional Development Plan, formally evaluate its impact, and take corrective action, if necessary.

From a development perspective, a successful leadership career (and rich life) is not about just putting in your time and moving up the organization chart. Success is more about navigating a variety of challenging and diverse experiences, learning from those experiences and then applying the lessons of those experiences to new challenges. From an Intentional Development perspective, we could almost express it as a Career Success Formula (with apologies to any reader who is a real mathematician!) shown in Figure 6.4.

Whereas, career success is the sum total of all Key Developmental Experiences (KDEs) throughout a career in which Key Leadership Competencies (KLCs) are called upon. Learning from the experiences is enhanced by regular reflection (Rfn) and feedback (Fdk). The benefits

$$\sum_{n=1}^{\infty}\left\{\left(KDE + KLC + Rfn + Fdk\right)^{LA}\right\}$$

FIGURE 6.4
The career success formula.

of the learning experience are heightened by an individual's ability to learn from experience and apply the learning in new situations (Learning Agility, LA).

When we coach executives or work with Talent Management leaders to implement Intentional Leadership Development, we highlight the following key components of Connect It.

Career Transitions: Let Go, Preserve, and Add On

One of the paradigm breaking results from the Center for Creative Leadership's Lesson of Experience (McCall et al., 1988) research was that when a leader's career runs off the tracks, or derails, it was often due *not* to a weakness or lack of a critical skill but the *overuse* of an existing strength or continuing to use a strength that is no longer applicable. We highlight this development challenge in the See It stage of Intentional Development; a leader may want to focus on reducing the dependence or use of a particular skill or balance for it with another skill in his development plan. These mis- and overuses often stem from earlier career experiences in which an individual was recognized or rewarded for a particular strength, a strength that is no longer effective in a newer higher-level role.

We like to highlight the model in Figure 6.5 in the Connect It stage, a model described by Arthur Freedman in 2011. The model emphasizes that at each career transition, effective leaders/learners must reinvent themselves. There are skills that must be added on to their repertoire, there are existing skills that must be preserved, and, maybe most importantly, there are vestigial skills that must be let go. Freedman referred to this as the Transformation Trilogy.

FIGURE 6.5
A successful career means changing at key transitions (Freedman, 2011).

For coaches and talent management leaders, it is important to help leaders understand when they are making a key transition and to help the learner to identify what the new role demands in terms of skills to add on and preserve and what is no longer demanded that can be let go (named in terms of competencies).

Career Development vs Development for a Position

The Connect It phase centers on building a longer-term perspective for leaders—a perspective on the individual's overall career and not just performance in a current role. For leaders, the discussion turns to what experiences to keep an eye out for or to aspire to that would enhance their portfolio of competencies. As a leader, once you understand the foundation of Intentional Development, you can then use the concepts to build a career—a career as moving *among* a series of developmental experiences and not just moving *up*. For many of the initial Intentional Development implementations, our focus is often on helping leaders learn from the challenges they are experiencing in a current role. Once that's under their belts, they can then be more *intentional* about what the next step or move would best enhance their career. That's the great feature of the link between competencies and developmental experiences: you can maximize the critical competencies you develop from an existing experience, or you can select experiences that can fill the gaps in your competency portfolio.

The Organization Building Blocks for Connect It

From an organization's perspective, the foundation for the Connect It stage of Intentional Development has two important building blocks— Competency Flow and Curating Experiences.

Competency Flow

First, there should be clear logic and flow among the competencies as the complexity and breadth of leadership roles increases. Many organizations create one set of competencies and assume they are the right competencies for everyone at all levels. Research (Brousseau, Driver, Hourihan, & Larsson, 2006) and our experience indicate otherwise. As described in *The Leadership Pipeline* (Charan et al., 2001), the demands of leadership jobs change significantly across different leadership levels in terms of complexity, breadth, time span of focus, and types and numbers of stake-holders. Managing an entire enterprise is very different from managing yourself or a single team. Subsequently, the competencies required for success are different.

We have found that it is not common for companies to initially build competency models that highlight what competencies to let go of at key transitions. This usually occurs later as their application of Strategy-Driven Leadership Development matures. Another way to address the importance of letting go is to make sure that any multi-rater (360) feedback assessments not only rate a leader on the competencies but to also provide the opportunity to rate the importance of a skill for the role and to identify potential competency overuses. This can be particularly effective if the 360 is used to help a new leader to onboard to a transitional role.

Curating Experiences

Another organizational building block for Connect It (and See It) is based on pinpointing the roles/projects/assignments that are significantly developmental within an organization and assigning leaders (identified in a robust talent assessment process) to those experiences that would benefit the most from those experiences (High Potentials, Emerging Leaders, Adaptable Professionals and/or maybe High Professionals or Key Performers that need to be tested). Not surprisingly, the roles that build strategy-critical competencies tend to be the ones that are strategic, pivotal, and/or scarce.

A whole science (and maybe art) has built up around this approach to matching aspiring leaders to specific experiences that will develop or enhance strategy-critical competencies. It's called "Assignmentology." We first learned about Assignmentology at the Center for Creative Leadership in a program entitled "Developing Successful Executives," presented by Bob Eichinger (then with Pepsico) and Mike Lombardo (then with CCL). We know that leadership skills are developed from navigating challenging experiences. Eichinger and Lombardo described how it is possible to identify what key competencies are required for success in different business challenges and then place leaders who need to develop or enhance those specific skills in those roles or assignments.

Managing the placement of talent into mission-critical roles for development purposes is not typically something that organizations are ready to do early in their Strategy-Driven Leadership Development journey.

We usually begin the discussion around Assignmentology once a critical mass of leaders has become comfortable with the science behind competencies, has built their skills in assessing talent, and has experienced the Intentional Development process. Once this tipping point is reached, you can then start cataloging roles and assignments that demand certain skills. Then, an important part of establishing a Strategic Talent Plan at the conclusion of a talent review would be *intentionally* matching agile leaders with confirmed development opportunities to targeted assignments and supporting them with an Intentional Development Plan. We find that the placement of talent is not typically based on who will benefit the most from the assignment but who can "hit the ground running." We like to emphasize that the better perspective is to place leaders who are ready to "hit the ground developing".

7

Coaching Reimagined: Building Intentional Coaching

Vince Lombardi, the venerable coach of the Green Bay Packers, always started off training camp by having all of his players sitting around him on the field. He would hold up a football and state to his team, "Gentlemen, this is a football." His belief was that every year his players needed to start anew to learn the basic and core skills of their profession. He must have done something right, as he and his Packers won three NFL championships and the first two Super Bowls.

Many managers make an assumption that their employees come to work generally well formed and ready for action. Sure, they may need some refinement or specific skill development, but we expect them to perform at high levels with minimum support or oversight. And interestingly, as people move up and the tasks become more complex, we have even greater expectations that they should know how to do everything even better.

In truth, however, most people come to work prepared to operate within their technical sphere of expertise. Engineers know how to engineer, social workers know how to counsel others, and salespeople can go out there and pitch a widget. These technical skills, as important as they are, only represent a small sample of the skills required for business success. As we've discussed, strategy-critical competencies play a more vital role across a range of leadership levels and are very different than technical expertise.

Managers and leaders want to help develop and grow their employees, and we've identified four different approaches that are used but typically not recognized as development methods. These are:

- Mentoring: Mentoring is a favorite development strategy for many leaders as it involves imparting wisdom. The mentor/mentee relationship is conducted over a cup of coffee with understandings and ideas shared, much like an academic experience where high ideals and best of world thinking are imparted to younger and thirsty up-and-coming leaders.

- Coaching: Coaching involves listening, asking questions, framing discussions, and driving change and growth in a structured and collaborative manner. The process advances the coachee's professional and personal growth in service to the success of the business. The task of coaching is more complex, perhaps, than the other development approaches in that it requires focused time to advance not just solution seeking but also improving the way the coachee thinks about how he or she approaches his/her work.
- Peering: In a peer situation, everyone is equal. There is much value for today's managers to see their direct reports as trusted colleagues with good ideas who bring their energy and commitment to the discussion. Peering helps level the playing field and builds a trusting relationship that allows for a free exchange of information. Given expectations of today's workers, this approach, as a development opportunity, provides the environment to strengthen connections so that colleagues experience mutuality in their relationships, an often overlooked but highly valuable workplace process.
- Bossing: Often times, it is necessary to just tell people what to do, and this form of development ensures that proper guidelines are adhered to and that learning occurs in a structured and efficient manner. Of course, some people will say that bossing as a development strategy is probably overdone, but if it is accomplished in an intentional and mindful manner, then it becomes a powerful tool in ensuring that procedures and actions taken adhere to organizational policies and expectations.

While all these modes of developmental support are beneficial and important, we'll be focusing the rest of this chapter on our notion that coaching needs to be a planned and thoughtful process that focuses on improving business outcomes by dramatically improving the specific skills required by leaders to achieve that goal. We call that approach "Intentional Coaching."

For those people who are doing coaching, being a skilled coach is not a "nice to have" but is instead a "must have" skill. In a 2010 *Harvard Business Review* article, researchers Jeanne Meister and Karie Willyerd surveyed 2,200 early career professionals about their expectations from the workplace and their managers. They stated that from their boss, they wanted

- Help in navigating their career
- Straight feedback
- Mentoring and coaching
- Sponsorship for formal development programs

To add some fuel to the fire of coaching as an essential leadership skill, Google's multi-year Project Oxygen, conducted by their People Innovation Lab, identified that being an effective coach was at the top of the list in what it takes to be an effective manager (Garvin, 2013).

On the other side of the equation is recognition that many business managers and leaders lack basic coaching skills themselves. A 2017 *Financial Times* survey found that while businesses were looking to recruit new MBAs who had coaching skills, most of these companies found that it is one of the hardest skills to recruit for in potential employees. In addition, many managers say they are not comfortable or skilled in providing coaching to their employees, and it may show up in terms of how frequently coaching is offered (Moules & Nilsson, 2017). In a 2015 study, entitled *Performance Management Research*, only 55% of employees indicated that they receive regular feedback or performance coaching (Jacobs, 2015). These findings are not surprising given that most companies, while recognizing the importance of developing leaders, fail to give proper time and energy to considering the skills required to achieve that outcome.

On an anecdotal basis, we find that many leaders love the idea of coaching others and many see it as a built-in skill. Perhaps they are modeling their behavior after a sports coach they've played for, a university professor who took a special interest in their studies, or even a manager who provided some good career advice along the way. They've lumped together the elements of the four ways to develop people we describe above and see it all as coaching. After all, seeing people grow and gain skills, experience success, and feel like one has made a contribution to the professional (or personal) growth of others is very satisfying experience. Yet, in today's more sophisticated workplace we need a more targeted and process-driven approach so that time spent is well spent and genuine efforts yield measurable results. That is why we think that our Intentional Coaching model, which can be easily described and enacted, provides a simple and actionable approach to implementing Strategy-Driven Leadership Development. So let's proceed to take a look at Intentional Coaching. For Coach Lombardi, that meant starting at the beginning, so perhaps that is not a bad place to begin.

A. WHAT IS INTENTIONAL LEADERSHIP COACHING?

We've already defined coaching as a collaborative process that involves questioning, framing discussions, active listening, and driving change to create a pathway for professional growth and business success. Intentional Coaching takes that process further by providing a structured process that helps both the coach and coachee to understand the steps that facilitate a learner's (or learners in a cohort) movement through the Intentional Development Process—Frame It, See It, Own It, and Connect It. Of course, the process is collaborative in that the coach needs to be well founded in the research and concepts behind our strategy-driven approach to development, and the learner needs to be committed to the change process, which means that s/he will have to do things differently

It took us a few rounds to refine Intentional Coaching, but we initially were clear on several aspects. First is that coaching is a collaborative process focused on building a growth mindset. Being a collaborative experience means that both parties see the coaching process as mutually beneficial and one that shares ideas, explores new perspectives, and tests and tries out actions to see how they will work. Second is that all coaching should follow the 70–20–10 model of development. We want it to focus on real-time activities that provide real time feedback for the coachee. Third is that we want the coaching experiences to be built into the work and not bolted on as something extra to do. We know

THE POWER OF MANAGERIAL COURAGE

We were coaching a new upper level leader, who we will call Susan, who was telling us that she did not think she had time for development even though her manager told her she needed to. During a talent evaluation, her peers and boss recognized that she needed to terminate one of her long-standing employees for poor performance. After doing an Impact Map, she saw how this employee was impeding progress for her group and she agree with the termination. However, she told us she never did terminations well. We talked about leadership skills such as making tough people calls and managerial courage, and her manager walked her through the process of the termination along with some role playing. After it was over (and she did well) we asked her if she now saw how she could build in her development in rather than bolt it on. Her response: "I now see that opportunities for growth are in front of me every day."

our coaching model is failing if we hear from people, "I just don't have any time to work on development."

As we've engaged in delivering executive and leadership coaching in the workplace, we've seen an incredible turnaround in how people view coaching. It no longer is something you receive only when you are in trouble but is something that is now viewed as rewarding and career enhancing.

Yet it unfortunately remains for many people as something that is done during an annual review rather than something that occurs all year round. Our intentional coaching model is designed to change that perception and action so that any leader within an organization can use everyday workplace activities as opportunities for learning.

B. KEYS TO EFFECTIVE INTENTIONAL COACHING

There are clear benefits to a properly designed and implemented coaching experience. But, as with any developmental tool, there are keys to assuring a return on the time and money invested.

Coaching Is Focused on the Right Players

As we outlined in the Development Strategies Matrix, the best opportunity to achieve a payback to a coaching investment is with high potential or high professional employees, managers, and executives. Helping these learning agile employees at critical junctures in their careers usually has a quick and sustained payoff. Coaching challenges are different based on the employee's career level, but the objectives are similar—tweaking an already top performer or assuring that a high potential employee achieves that potential.

Figure 7.1 highlights which Development Strategies would be best supported by a professional coach—High Potentials, Adaptable Professionals, and Emerging Leaders. Depending on the level and criticalness of the learner's role, it may be effective to employ a coach to help a Diamond in the Rough to achieve his/her perceived potential. Likewise, there may be some Key Performers that, with the right Intentional Development Plan and coaching support, may improve their performance and/or potential. We often find this as a way which organizations uncover some "hidden talent gems" that can help increase overall leadership capacity. Coaching

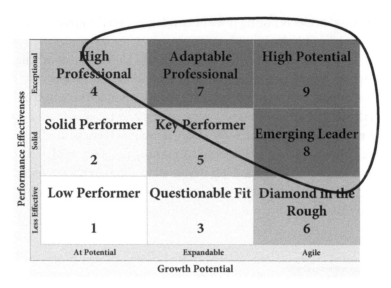

FIGURE 7.1

Development strategies and best coaching applications.

for High Professionals may focus on helping them build their networks and networking skills or building their ability to mentor others and pass on their unique knowledge or relationships.

We have often seen coaches being employed to assist an employee that is perceived to be "broken" and the goal of the coaching assignment is to "fix" them (Low Performers). The coach is imposed on the reluctant employee, a lot of personal and organization stress ensues, and the usual result is that the employee disengages (retires on the job, quits) or ends up being terminated (which probably should have been the action in the first place). Coaching to improve performance has a higher probability of success for those seen as Questionable Fits or Diamonds in the Rough who are perceived at least as having some potential.

This is not the optimum use of coaching resources. The "broken" employee most likely has a history of blocking or barely supporting critical change initiatives, has a reputation of being a "people eater," or has a history of poor performance. They are typically low learning-agile people who have already derailed and getting them back on track is a high-risk, low-benefit proposition. Employing regular talent reviews and planning follow-up based on the Development Strategies Matrix helps avoid this situation.

Our perspective—if you have $100 to spend on development and want to get the biggest bang for your buck, spend $70 on increasing the impact of High Potentials, Adaptable Professionals, Emerging Leaders, and High Professionals, $20 on maintaining the capability and competitiveness of Solid and Key Performers, and $10 compassionately getting the "show stoppers" out of the way.

Coaching Is Applied in the Right Situations

Coaching is not a panacea for all development needs nor should coaching be employed outside of an Intentional Development outline. Coaching is best employed where specific improvement goals and clear benefits have been identified in an Impact Map and outlined in a Learning Map and not just "finishing school" kinds of challenges. Transitions, preventing derailment, and addressing blind spots are also some specific opportunities for coaching.

Transitions

Starting a new position with a company, moving to a new role within a company, or taking on new, broader responsibilities are critical junctures in a career. As we've described, it is a time to let go of competencies that are no longer relevant, to sustain competencies that are still key, and to add new competencies that now come into play. Providing a coach to design and facilitate Intentional Development during a transition process can have substantial benefit. These are stressful times when previous weaknesses become highlighted or overused strengths come to bear. An effective coach can help assure a smooth transition for the employee and their new stakeholders, create confidence by building quick wins, and create effective feedback so that learning gets locked in.

Preventing Derailment

As we've described, the causes of career derailment are numerous. It may be related to not being able to let go of previous responsibilities and an associated failure to delegate. A lack if personal insight may lead to an inability to adapt to changes and a persistence to do things the way "they've always been done." Even high potential employees can stumble at times, most often due to an overuse or overdependence on a strength. Coaching can prevent derailment by helping the employee identify potential failure modes and by developing specific preventive and contingent action plans to address them.

Addressing a Blind Spot

This is the more typical coaching situation. We've all seen it—an otherwise promising employee lacks a specific skill or exhibits behaviors that impact his relationships with others. These can be derailers but are more often opportunities to coach high potential, high performers with a blind spot or untested competency into well-rounded, great performers.

There are specific situations where coaching should not be used.

- Unethical or illegal behavior is the issue and the coach is really meant to be a detective or a buffer between the employee and others.
- When there's a long-term performance problem that should be immediately confronted by the employee's manager or when the coach is expected to create a case to terminate the employee.
- When somebody has decided that the employee needs a coach, without the buy-in of the employee and the employee's manager or without the guidance from an objective talent review.
- When the employee has psychological problems that should be addressed by a medical professional, or at least a coach with clinical experience. A good coach should be able to quickly identify the warning signs and make a referral. Coaching is not therapy and should not be an intervention going after deep psychological issues.

The Right Skills for Being an Intentional Coach

Not every leader or manager is cut out to be a great coach. Some lack the emotional intelligence to read and understand people well. Others may not have the natural inclination to want to help others. Some may lack the self-management to understand how their efforts can drive business success. Regardless, we don't have to have everyone be a great coach, but instead be intentional about looking for the opportunities to help the business succeed by developing others. In the same way that we believe we can develop great employees through coaching, we've seen time and again that just about every leader can be a better coach by paying attention to these six key skills of Intentional Coaching.

- **Business Acumen**: Coaches need to be good businesspeople first. They must be able to understand the market challenges and business priorities that their clients face. They need to be able to understand and use the language of executives. This is particularly important when it comes to translating business goals and priorities (Why is this important?) into targeted development opportunities (How does this happen?) in an Impact Map.
- **Development Expertise:** Effective coaches need to understand the foundation concepts of Intentional Development such as adult learning, the neuroscience of leadership, the 70–20–10 development model, behavior-based competencies, learning agility, and the It Formula.
- **A Commitment to Development**: Effective coaches want to help others succeed and with that success, comes more success for themselves and their business
- **Being in the Moment**: The best Intentional Coaches are often usually pretty good at seeing what is happening in real time and are able to respond accordingly. They understand and take an interest in individual and group processes and can guide individuals and teams through effective questioning, analysis, planning, and decision-making processes. They can read individual and group dynamics quickly and respond appropriately. This skill is particularly important for building the coaching opportunities in at the moments when they occur rather than waiting until our "regularly scheduled meeting time."

THE BEST COACH

We were meeting with a senior leader at a financial institution who told us he was the best coach they had. When we asked him to explain more about that, he told us that he not only had his employees ask him frequent questions about their development, but his colleagues would often ask him to meet with their staff to help coach their development. He said that his business success coupled with his experiences as a volunteer with community leadership program had prepared him to understand others and how to help them with their own leadership development skills. Given his track record and his commitment to developing his coaching skills, we had to agree with him.

- **Credibility**: The best coaches have had business or other experience that can allow them to maintain a perspective. They build credibility quickly by showing that they have "been there, done that." Credibility does not come from a bag of tricks. Beware of a coach who only has a one-size-fits-all technique or instrument that they apply in any situation.
- **Interpersonal Savvy**: These leaders like other people and gain a sense of delight out of helping others use their most effective skills for success. They are able to relate to all kinds of people, build rapport and relationships easily, listen, use tact and diplomacy, and deal with tense situations with a good measure of grace and success. This skill is important because under challenging times, people tend to revert to their least effective strategy and having a coaching mindset that allows you to see people's strengths rather than their weaknesses provides a forum for bringing about changes in real time.
- **Resilience**: Coaching is often about seeing the failures and losses and turning them around so that a positive learning experience can occur. Overcoming the inertia of the negativity bias can be difficult and expanding the silver lining so that it becomes a golden path to learning is an important skill for the best intentional coach

Coaching Is Integrated

As we have emphasized in our strategy-driven approach, Intentional Development and, therefore, Intentional Coaching should not be designed and implemented outside of the company's strategy, culture, and existing business processes. The business case for coaching should develop from the company's strategy. The Strategy-Driven Leadership Development model highlights that the need for coaching is best identified from a talent review, driven by the strategic needs of the business. Coaching should not be a one-off tool. It should be applied where it can best improve a strategic capability of the organization.

If your organization already has a well-designed competency-based talent system, those terms and concepts should be well understood and then integrated into the coaching by the manager/coach. Many managers who enjoy coaching typically developed their approach through their own experiences and lack an organizing schema for coaching. Bringing them back to the leadership and values competencies established by the organization helps remind them of the strategies and also provides the key concepts that want to be used in coaching.

The process of how new behaviors and capabilities are developed and put into play is better understood and enacted by using the Impact Map and the Intentional Development Tracker. One of the pleasant surprises that we've seen come out of this work is that since it is so easily understood, leaders and their followers put it into play to develop new sets of behaviors for changing circumstances. As the process becomes more attuned for the leader, s/he wants to use it with his/her direct reports, and so the process can organically be cascaded on down through the organization. While it is best sustained through the company's existing performance planning and review process, the reality is that unless leaders, managers, and coachees see it as working, it won't be used. The employee's manager can help reinforce the work by returning to the process when leadership development opportunities arise.

C. LEADERS AS COACHES: REIMAGINING YOUR APPROACH

In Intentional Coaching, we want to reinforce the notion that coaching (like Intentional Development overall) is "built in and not bolted on." The opportunities to coach members of the team occur every day and managers can look to bring these opportunities into play on a regular basis.

Here are 10 ways that any manager can retool his or her approach to coaching to be more *intentional*.

1. Clarity around objectives
2. Get the jump rather than respond
3. Use power appropriately
4. Focus on what matters most
5. Show Impact
6. Don't forget the team
7. Encourage peer mentorship
8. Talk about career planning
9. Encourage diversity of people, ideas, and actions
10. Remember your development

Clarity around objectives: I recently had a call with a senior leader about coaching one of his directs and he asked what they should be focused on. When I asked him why he was coaching this person, he responded that he needed to have some of his "sharp edges" smoothed out a bit. I followed up by asking him how his employee's behavior impacted business? That helped clarify the focus as he could now think of two or three events where this person's behaviors threw a discussion off focus or where he said something that did not facilitate a discussion with a customer. I reinforced the importance of matching the needs of the business against the skills of the employee and nail down objectives that would most benefit the firm and the employee. This is the perfect place to work jointly on developing an Impact Map. The step-by-step questions that we answer in building a Map are an effective outline for the initial coaching conversation.

Get the jump rather than respond: There is always an opportunity to respond to issues and concerns as you see them bubble up. Conducting action planning and regular follow-up from talent reviews provides the opportunity to be proactive on talent development strategies and not just respond. Additionally, your employee will be grateful that you have initiated the development discussion because s/he would have been thinking about it. Remember that as the manager you are enabling the discussion and not responding to it.

THE UNCOACHABLE EMPLOYEE

Mark thought that he was just getting too much work from his boss. He had responsibility for event management, membership growth, and now he was being asked to take on a new employee who would do some work for him but would also be helping other managers.

In our coaching session, he spent a lot of time explaining why he couldn't get all these things done and be successful. When we talked about a planning process to improve his organizational skills (which he sorely needed to do), he would whip open his computer to show the spreadsheet he preferred to use rather than his firm's Project Planning software. When tracking contacts, he shared his own internal database system he insisted was better than the Salesforce.com app that the company used for tracking client information. "That is just way more complex than we need" he would tell me.

The only thing that seemed to work with Mark was to give him one-off suggestions on how to problem solve specific issues, which was helpful but didn't meet his needs in the long term, or the needs of the firm.

After three months of coaching with Mark, my recommendation to him and to his boss was that he would probably not grow into his role the way they needed him to achieve.

Not surprisingly he disagreed with my assessment and recommitted to work on his approaches but his boss already saw the writing on the wall, and Mark was soon replaced, not just because of his performance but because he was so unwilling to grow from the coaching experience.

Meeting the employee's needs: Manager-coaches want to make certain that they remember that the needs of the coachee are also critical and must also be taken into consideration as they are helped to develop. As we emphasize in the talent review, recognizing the importance of the coachee's aspirations is critically important to the success of his/her development. As the manager is coaching, s/he always wants to couch development as something that is good for the firm and for them. Again, the Impact Map and the Personal Business Cases serve as great vehicles to guide the discussion and capture specifics.

Focus on the vital few: It is not unusual for people to want to work on 8–10 different skills but focusing on **what matters most** will get people advancing in their development. As we've said before, learning new skills is not hard to do and can be advanced in just a short period of time. The key is that people have the chance to focus on the key actions, which is identified in the See It stage of Intentional Development.

Show Impact: Like our coaching client, Susan, in the above example demonstrated, small steps in building her understanding of why the termination was important along with the specific skill building of managerial courage and making tough people calls made the tactical role playing exercise an afterthought. In most coaching situations, people would say that learning how to terminate was the big skill learned. For our model, we would say that this leader gaining more understanding of why and then how to make a termination (making tough people calls and managerial courage) was far and away a more important learning.

Don't forget the team: Development is not just about the individual but also the team. As we've said before, while many people work well on a team, others do not, and team-effectiveness competencies can be delineated on the Impact Map and built in the Intentional Development Plan in such a way that the team understands and builds the skills of teamwork for their success. Particularly when it comes to a significant change in strategy with far-reaching impacts, we have found it beneficial to do Intentional Development on strategy-critical competencies for whole teams.

Encourage Peer Mentorship: While you are bringing the team together to learn about team skills, also examine ways to get team members to support each other's development. Early in Richard's career, the COO of his company held regular bi-weekly luncheon meetings with his direct reports. Team members would informally gather in his office and would discuss everything from work challenges to how to balance time with team members who had young families. If someone returned from a conference, s/he would share his/her findings and the team would discuss how we could apply it to our workplace. We called the meeting "breadcrumbs" because we'd get so enthralled in the topics that we forgot to clean up our lunch mess and frequently left those reminders on his table or floor.

Talk about career planning: One of the great challenges facing younger employees is where and how they will move up in organizations. Companies are becoming flatter, senior leaders are hanging around longer as they find renewed meaning in their work, and younger employees may feel there is no place to go . . . and they may be right. Your HR department may have a career mapping process for employees, but probably not. Having that discussion with team member demonstrates a genuine interest in their work as they think through their next opportunities, which may include one where they can deliver more value for the organization. The Connect It step of the Intentional Development Process provides the opportunity to talk about development from a career perspective.

Encourage diversity of people, ideas, and actions: Diversity will always need to focus on how we can address the needs of underrepresented people in the workplace. Research shows that the more diverse opinions and people with experiences we have in our world, the better the solutions we arrive at. Diversity also means that we create different experiences for people to engage in as part of their development. Getting engaged in civic activities, participating in leadership programs that contribute to the larger economic engine in the community, for example, creates new vistas for anyone in the workplace. Not all 70s or Developmental Experiences in a Development Action Plan need to come from within the organization.

Remember your development: Development is not just for your employees but also for the leader. What can you, as a leader, do differently and learn to do in a new way that will stretch you? Modeling a behavior can be an important 20 in a Development Action Plan. For example, turning a coaching session into a peerage session by soliciting ideas on new uses for social media not only helps inform the leader's understanding but also empowers the employee to see ways to contribute to his/her manager's growth.

At its simplest, intentional coaching is about having a planned and an effective approach to development. Most leadership development coaching is not only haphazard but lacks a focus that emphasizes business outcomes while respecting the needs of the person being coached. With a few simple tools and basic coaching competencies, leaders can begin transforming everyday opportunities into learning experiences—that is, if your employee is open to learning.

The Coachable Employee

Would it surprise you if I told you that some of your team members don't want coaching or advice? They will probably listen to directions (bossing) but that does not mean they think they have much to learn from anyone about their development.

They may have a point, as many employees' experience with professional development has not been a fruitful journey to career advancement. Implementing the Intentional Development model will demonstrate that the firm is committed to their development and that they can actually learn something that will benefit them professionally and in their career. Some employees will embrace the process and others will view it as an intrusion of their time when they are trying to get their work done.

Every organization and every team has some people who are uncoachable. They may be seen as people who are moderately or fully disengaged from work and are merely focused on surviving in their role. Their lack of commitment is further demonstrated by their unwillingness to learn or to look for opportunities for growth.

It's not hard to tell the difference between a coachable employee and an uncoachable employee, but the unfortunate truth is that many times leaders will not know the difference because they are not focused on the right person to coach as we discussed previously. Too much time is typically spent on the low performer who probably doesn't want or see the benefit of spending more time than s/he needs trying to fix an employee who can't be fixed instead of focusing on developing that employee who would benefit the most from getting some great coaching.

Here are five ways that you can determine whether your employee is coachable and then you can figure out if they are not coachable:

First and foremost, they learn more from experience; they are learning agile: DeRue and Wellman (2009) found that individuals with a high learning orientation maintain their focus on learning from developmental challenges (the 70s of the 70–20–10 development model) and therefore, gain more leadership skills from developmental opportunities.

They want to be challenged and to learn and grow (Results Agility): Carol Dweck identified that some people have a growth mindset and some people have a fixed mindset. People with a growth mindset understand that we all continue to learn and grow while those people with a fixed mindset question whether they can and should learn new things (think of it as "we've always done it this way"). From the beginning of the time your employee joined your organization, you can tell whether they thirst to learn new things or whether they are just riding it out. Coachable employees want to learn more and then they do more.

They welcome feedback (People Agility, Self-Awareness): For many people receiving any kind of feedback is difficult. Good news embarrasses them and bad news dismays them. Coachable employees take both kinds of feedback in a mature and appropriate manner. They are good listeners, validate and point out their areas of agreement, and are not afraid to ask for clarification and even push back when they disagree with a finding (good or bad). For example, one of my coaching clients told her manager that she could not take the credit for the success of a project but that instead her second-in-command had really led the initiative and deserved the accolades. Beware however, that most folks are not comfortable getting bad news and may need some time to learn the skills associated with receiving this kind of information (such as not getting defensive, seeing the information as a gift rather than a threat, and learning how to take the information and use it to develop a new plan).

They take responsibility (Results Agility): How refreshing is it to have someone take responsibility when things go bad. "I screwed it up" would be what you would want to hear from someone when a project misses a timeframe or communication mistakes lead to a failure of collaboration. In a 2015 *Harvard Business Review* article Jim Whitehurst, CEO of Red Hat, told the story of putting a product out to market before it was ready and then pulling it back and delaying a new release for over a year. Both the decision making and lack of communication left his team with a sour taste in their mouths. To his credit, Whitehurst took the blame and explained what happened and how he would do it differently in the future. It's easy for people to forgive and move on when they see others take accountability

They are open to looking at things in new ways (Change Agility): Today's workplace demands agility as a core competency. Things change too quickly to have a single point of view, and coachable employees want to explore different options. In our executive coaching work we want people to consider four to five different approaches to solving a problem, not two. Getting out of binary thinking allows your team members to see more possibilities. On the flip side, that message of "Why should we change?" is a hallmark of stagnant thinking

Just Do It (People Agility, Results Agility): Even though we are psychologists, we don't get hung up on people's personalities. We are focused on behaviors and the behaviors that are necessary for professional and personal success. We've said before that changes don't take years but months. Research shows that a committed effort to modify or change a behavior takes about two to three months. Perfection is not the goal; progress is what we are looking for, and a coachable employee's effort will show results by then. I had a client leader once whose feedback from his team indicated that he needed to come to meetings on time and with an agenda that was sent out at least 24 hours in advance. After giving him templates for agendas and strategies around time management, he persisted in explaining to all that it would probably take him a year to change his behavior in regard to this organizational process. Not surprisingly, he had other issues around his own personal management at work and as a result, he was eventually demoted to a position where he served as an individual contributor and not as a leader.

Coaching your employees is a powerful way of assessing your team members' commitment and engagement to their work. Who wouldn't want the opportunity to grow and improve in their work? But surprisingly many people either don't know how or don't want to take in feedback. As a manager you can use this interaction as more proof to the good or bad side about your employees. For sure, you want to work with them and even explain why their development is important for their career and that how your best intentions are there to support them. Yet, you can't do any more for them than they want to do for themselves, so make sure their development isn't more important to you than to them.

If you find team members who are uncoachable, let them know the behaviors they are demonstrating that give you that message and what it is that you expect from them. If that works, great; if it doesn't, then put your energy into those who want to grow and watch carefully how those who are uncoachable have their career play out.

It won't be pretty.

"Hey Coach, Put Me In" . . . "Hey Coach, What Should I Do Now?"

If only it were that easy that your team members came to you to ask what they should do and even if they did and since you are already a great coach, your answer would be . . .

Most of your team members don't have much more of an idea about their development than you did (before you read this book). They probably assumed that you or someone over in HR had prepared a career path plan that would be looking out for them. Perhaps they believed their annual review would not only highlight their past year achievements but would somehow describe how their future would play out in this company.

You can and probably want to change that perspective for them and not just for your younger professional colleagues but also for the seasoned professionals in your group. The younger ones will remember you as a leader who transformed their life and the older ones will thank you for bringing new meaning and excitement to their career.

Intentionality is defined as being something that is done on purpose and in a deliberate manner. Intentionality is not about asserting your behavior or even about being decisive. It is as much of a thought process as it is an action process in that anyone who is being intentional does so in a planned and intended manner.

Much of our workdays are not driven with intention, despite our best efforts. They happen more in our firefighting role (we never support the idea that people are involved in fire drills; they are actually in the midst of firefighting). Most leaders beg to have more time to deliberate on how things work but rarely do, but even real firefighters must and do take time to review their actions and plan for the next event. The fire chief will meet with all the firefighters in the station and review their performance during an emergency and provide recommendations for improvement. After all, their lives depend on it.

8

Gauging Impact: Evaluating Intentional Development and Talent Analytics

Value does not come merely from the exposure to the training or the acquisition of a new capability. . . . Value comes from the performance changes that the training actually leads to.

—**Mooney & Brinkerhoff**, *2008, p. 36*

For learners, the Impact Map from the Frame It stage of Intentional Development sets the context for their development. For talent management practitioners, an Impact Map also provides the foundation for a simple but very effective tool to evaluate the success of any development initiative. The evaluation tool, called the Success Case Evaluation Method, has been described elsewhere (Mooney & Brinkerhoff, 2008; Brinkerhoff, 2006), so we'll just provide a quick overview as it relates to Intentional Leadership Development. The Success Case method gathers evidence on three basic questions after the implementation of a development plan.

1. Is the new competency being used? If not, why not?
2. When the new competency is being used, what good does it do? Are they seeing the results outlined in the Impact Map?
3. What would it take to get more value from the using the competency in new and different challenges or roles?

Figure 8.1 diagrams the steps to take to evaluate the impact of an Intentional Development Plan. We have used this to evaluate individual Plans as well as the development of whole Development Cohorts. It can be completed as a written survey or by interviews.

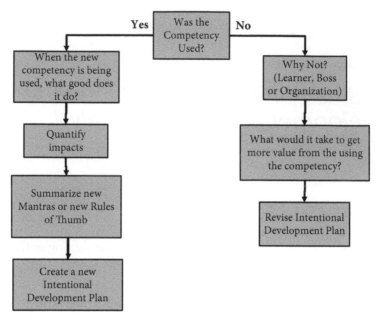

FIGURE 8.1
Connect it: intentional development evaluation.

The timing of when to complete the evaluation depends a lot on the type of intervention. If we are working with a Development Cohort, the cadence of development usually continues over a specified period of time during which the cohort members have a variety of development experiences, get frequent feedback and coaching, and complete regular development tracking. The specified period of time is typically six months, so we may do an evaluation halfway through the cohort experience and at the end. If we are working with an individual learner, the timing of the evaluation depends on the milestones in their Development Action Plan when they have had enough development experiences and time to reflect.

We find that the Success Case Method provides a very rich and robust evaluation tool. Learners can easily describe times when they used new behaviors and can link specific outcomes to their application. This would not be the case if we did not use our unique approach employing an Impact Map, a Learning Map, a Development Action Plan based on the 70–20–10 model, and regular development tracking.

By the way, if your organization already has structured leadership development programs in place, the Success Case methodology is a great evaluation tool for those approaches as well. Start by creating an Impact Map that links the content of the program to business strategy (a good step to take anyway to make sure that the business case is there). And you would need to translate the content of the program to specific competencies or behavioral descriptions. The evaluation approach would then be the same as described above.

The Impact Map is also something that should be shared in advance with the learners attending the program so their time and effort is framed in a business context. The best person to review the Impact Map with the learner is not the trainer but the learner's manager. That way the manager is linked into the development process and can help with follow-up after the development program.

TALENT ANALYTICS USING THE DEVELOPMENT STRATEGIES MATRIX

It was once said that the universe can be described in a series of simple matrices. So it is with the Development Strategies Matrix that is generated from a talent review process, first outlined in Chapter 4 (but in this case now looking more like a typical 9 Box or Performance Potential Matrix, since it's not highlighting development.) For example, the matrix can be used to display the overall investment in talent by summarizing simple percentages of talent by cell (versus just listing names in the cells). This type of summary is a great place for you to begin a strategic talent discussion by asking the question, "Based on this picture of talent, does it look like that we will be capable of executing our strategy?" This will, of course, raise more questions than a single matrix can answer, but it often leads to additional slices of the talent data that can be more enlightening. The idea is to use the different slices of data to drive deeper discussions that lead to meaningful actions. See Figure 8.2.

Another perspective can be gained by summarizing total compensation in each of the cells as in Figure 8.3. We have clients who track the ratio of 4, 7, 8, 9's total compensation to 1, 3, 6's compensation over time (Figure 8.4) to see if they are making progress on their overall organization capability and investing their compensation dollars in the talent that would have the greatest return. Tracking total compensation by talent cell overtime also shows this capability trend.

Displaying average CompRatio (base salaries compared to market target) by cell, as shown in Figure 8.5, provides insight to how competitive

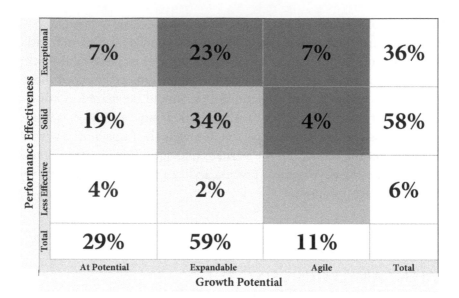

FIGURE 8.2
Talent distribution by cell. Do we have the talent we need to execute our strategy?

	At Potential	Expandable	Agile	Total
Exceptional	$382,071	$1,859,313	$2,048,926	$4,290,310
Solid	$1,487,147	$10,012,281	$1,227,405	$12,726,833
Less Effective	$524,626	$231,920		$756,546
Total	$2,393,844	$12,103,514	$3,276,331	$17,773,689

FIGURE 8.3
Total compensation by cell example. What does our overall investment in talent look like?

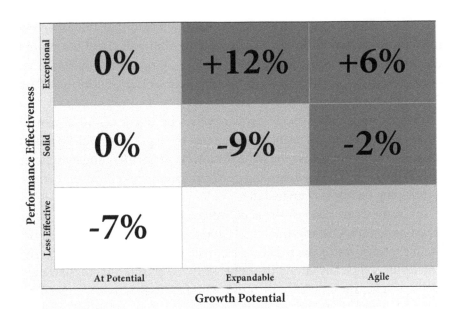

FIGURE 8.4
Total compensation time 1 versus time 2 example. Did we improve our talent investment and capability?

	At Potential	Expandable	Agile	Total
Exceptional	1.10	1.05	1.08	1.07
Solid	1.13	1.00	0.97	1.05
Less Effective	1.09	1.11		1.09
Total	1.12	1.02	1.03	1.06

Performance Effectiveness (vertical axis)
Growth Potential (horizontal axis)

FIGURE 8.5
CompRatio by talent cell example. Are we using compensation as an effective talent lever?

your key talent is against the market. We often find that the differences in CompRatio have little to do with the true ability of talent but more to do with service or time in role. Are you leveraging compensation to recognize contribution and to ward off turnover of High Performers/High Potentials?

Calculating simple average demographics by talent cell also provides another perspective on talent. In the actual example shown in Figure 8.6, longer service employees tend to be the lowest performers and least potential. Why? Are we not addressing performance issues effectively? Are we not helping people stay up to date?

The matrix can also test the effectiveness of other talent processes. For example, Figure 8.7 shows the actual data when we matched average performance ratings versus the results of a detailed and calibrated talent review. The result is typically always the same when we do this—there is no statistically significant difference in performance ratings across the matrix. In this and many other cases, performance review ratings are not aligned with other talent processes and do not really differentiate among performers—everyone is above average and the ratings do not reflect true contribution. As research often finds, typical performance ratings tell us nothing about the overall capability of our talent. There are just too many inherent biases.

We have also used the results of talent reviews to assess the effectiveness of hiring and promotion decisions. Figure 8.8 looks at recent hires and

FIGURE 8.6
Average service by cell example.

	At Potential	Expandable	Agile	Total
Exceptional	4.0	4.3	5.0	4.4
Solid	5.0	4.7	5.0	4.9
Less Effective	4.7	4.8		4.8
Total	4.9	4.6	5.0	4.8

(Performance Effectiveness vs. Growth Potential)

FIGURE 8.7
Average performance ratings by cell (5-point scale).

	At Potential	Expandable	Agile	Total
Exceptional	1	4	8	13
Solid	2	4	2	8
Less Effective	1	1	2	4
Total	4	9	12	25

(Performance Effectiveness vs. Growth Potential)

FIGURE 8.8
Talent distribution of recent hires and promotions example.

promotions. For this organization, they were missing something in some of their selection decisions or onboarding—a number of recently placed folks are not performing effectively.

The performance potential matrix can also be sliced and diced by the talent mix that is currently in mission-critical roles (pivotal, strategic, and/or scarce). This perspective creates a picture of risk and highlights where you have opportunities to improve productivity or value creation. The 9 Box in Figure 8.9 is from a small organization and shows 10 mission-critical positions filled with less-than-ideal talent. The capability of the talent in these roles needs to change or their growth strategy could be in jeopardy.

After an organization has several years of talent review data under their belt, you can also look at Time 1 versus Time 2 (or more) changes. For example, Figures 8.10 and 8.11 show an example of the overall Development

FIGURE 8.9
Talent matrix of strategy-critical roles.

FIGURE 8.10
Talent assessment at time 1.

Strategies Matrix from two talent reviews for the organization's top leadership positions (expressed as a percentage of total salaries). Time 1 (Figure 8.10) indicated a significant gap in overall capability particularly with Less Effective performers in pivotal roles. Time 2 (Figure 8.11) shows the results after two years of restructuring and development. A simple statistical test can be used to determine if the results are statistically significant.

Another useful comparison over time identifies the promotion rate by development strategy as shown in Figure 8.12—the percentage of people

Time 2

Performance Effectiveness	At Potential	Expandable	Agile	Total
Exceptional	7%	18%	11%	36%
Solid	11%	30%	10%	51%
Less Effective	10%	0%	0%	10%
Total	28%	48%	21%	

Growth Potential

FIGURE 8.11
Talent assessment at time 2.

Performance Effectiveness	At Potential	Expandable	Agile	Total
Exceptional	13%	31%	0%	23%
Solid	11%	20%	45%	19%
Less Effective	0%	17%	0%	8%
Total	11%	24%	31%	19%

Growth Potential

FIGURE 8.12
Promotion rate by development strategy.

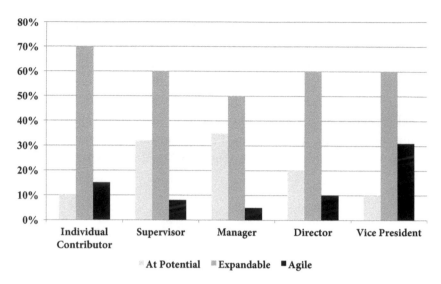

FIGURE 8.13
Talent pipeline capability.

who were promoted from time 1 to time 2 by Development Strategy. This analysis highlights whether the talent assessment information is actually used in making talent decisions. In this organization's case, there is a link between promotion and Growth Potential. However, no High Potentials were promoted.

Beyond displaying analytics in multiple slices of a matrix, we have also worked with our clients to develop and track several other key metrics. For example, you can create a picture of your overall talent pipeline by plotting Growth Potential by organizational level as in Figure 8.13. This organization has a potential gap in the pipeline at the manager level, with a low number of Agile players and a high number of managers who are At Potential. This highlights the need to look at what turnover or succession is on the horizon for key positions at the Director level and to determine if there are sufficient internal replacement candidates. Can some of the Expandable Managers be stretched and how long will it take them to be ready?

Additional talent analytics that have been proven to be valuable to track include the following.

- percentage of pivotal and strategic roles filled from within the organization (internal placement rate)
- percentage of pivotal and strategic roles with identified successors (bench strength)

- percentage of 4, 7, 8, 9s (from the Development Strategies Matrix) with Intentional Development Plans
- retention by Developmental Strategy

The data you gather in the talent review process is very rich. It is well worth your time and effort to build talent analytical tools and capability within your organization. Business intelligence software makes this task easier but we often do this robust analysis using simple spreadsheets and pivot tables.

9

The Strategy-Driven Leadership Development Journey

Where do you begin the journey to build Strategy-Driven Leadership Development (see Figure 9.1) into the DNA of your organization? The simple "duh" answer is with your organization's strategy. However, we have found that there can be several different "on-ramps" to the journey, depending on the talent history and culture of our clients' companies.

On-ramp 1: Talent Assessment: Surprisingly, we often start in the middle of the SDLD model (Leadership Capability) because of a client's need to better understand his/her full picture of his/her talent, particularly at the top levels of the organization. This is often due to an onslaught of potential retirements from key roles, because of rapid growth requiring an expansion and/or the Board of Directors asking questions about succession. Our approach to talent assessment allows for an objective evaluation and summary of talent capability that can be used to answer all of these challenges. We can also work backward in the SDLD model and use the behavioral data gathered in talent assessments to begin to build competency models based on the organization's best and brightest. We can then move back another step in the model and compare the draft

FIGURE 9.1
Strategy-driven leadership development.

models to their strategy and to further validate the critical competencies. (We can also confirm mission-critical roles through the process.) After we have, in a sense, filled in the blanks, we implement Intentional Development.

On-ramp 2: Upgrade existing leadership development: We are often asked to work with an organization because they have a Leadership Development program that is not having the desired impact. The program may often be based on targeted competencies (See It) but has missed the Frame It piece or is solely training-based and needs a dose of Own It and See It. This can be one of the quickest payoff on-ramps because, since the organization has some talent infrastructure in place, it requires more of an upgrade than a whole new system.

On-ramp 3: Build Competency Models: Most mid- to large-size organizations (for- and non-profit) typically have a well-defined and articulated strategies. With that in place, translating the strategy into strategy-critical competencies is a perfect on-ramp. We've found that beginning at this step is often a feather in the cap for the HR function because we provide an approach which allows them to interact with top leaders in a strategic fashion and to build a clear link from their work on talent to the success of the business.

On-ramp 3: Strategy Execution: Since we do strategy development, another typical on-ramp is a key part of deploying the strategy across the organization and at all levels. The work revolves around answering the questions "What kind of talent do we need for our strategy to be successful?" (Talent Demand) and "How many and what kind of talent do we presently have in the pipeline?" (Leadership Capability). This on-ramp takes the most work but is very efficient in linking talent and talent development to strategy.

On-ramp 4: We Got Nuthin': We had a client whose last strategic plan was years old and, therefore, basically non-existent. They also had no real talent management infrastructure nor any approach to leadership development besides some basic leadership skills training that wasn't being applied. Their competitive environment was changing rapidly. Their new CEO needed speed, so we started with talent assessment at just the top-level positions. Unfortunately, the picture of leadership capability was not very positive so we worked with the CEO and her most skilled leaders to compassionately change out the players, particularly in mission-critical roles. We could then go back and start the process of building skills for those in the pipeline so they had some bench with which to work. With the talent in place, they could then begin the work to define and build a new strategy to assure their future growth. This was like on-ramp #1 but much more accelerated.

On-ramp 5: Leadership Transitions: There may be no better time when employees are new to the organization or to a new leadership role. They are malleable about philosophy and values, usually have some time to orient themselves, and have probably taken time to assess and evaluate their life and career and where they want to go next. In addition, global research has highlighted that many organizations are not very good at assisting executives when they transition into VP level and above roles, even though a large majority of the executives felt that transition assistance made a major difference in the early impact (Byford, Watkins, & Triantogiannis, 2017). Creating the expectation among newly placed leaders that development is important fuels the organizational need for matching business needs with the right talent and helps the employee recognize where s/he may need to improve to meet those business requirements. Additionally, this on-ramp can be more easily accomplished through a formal "onboarding process" that is developed and directed in much the same way that an employee signs up for benefits or learns about ethics guidelines.

A. SDLD MATURITY: IT'S A JOURNEY, NOT A ONE-AND-DONE

Just as effective leadership development is not a one-and-done, implementing Strategy-Driven Leadership Development should not be envisioned and implemented as an HR program but as a core business process that can be improved and advanced over time. We've seen that the maturation steps are not exactly the same for every organization, due to differing strategic demands and to the differing on-ramps we described in the previous section. Beginning with the initial pain points, we like to work with our clients to develop a vision for their strategic talent practices that best encapsulates where they are headed as a business. We work jointly to define the current maturity level of their talent practices, to define how they envision the future, to specify what the payoff or benefits will be and to outline the rough steps it will take to get there—sort of an Impact Map on a grand scale. This discussion often starts with the folks in Talent Management but needs to spread quickly to Executive Management or whoever approves the plan and investment.

To help our client partners better analyze and visualize the maturity of their talent practices, we like to use a version of the People Capability Maturity

Model that came from work at Carnegie Mellon University's Software Engineering Institute. The People CMM was built to provide "a roadmap for implementing workforce practices that continuously improve the capability of an organization's workforce" (Curtis, Hefley, & Miller, 2001). It outlined five maturity levels that could be used to describe the maturity of any talent practice in ascending levels of value. We revised the five levels somewhat mostly because the original People CMM did not have a level where the capability was nonexistent . . . and we have worked with organizations whose leadership development practices and processes were nonexistent.

Here are descriptions of the five maturity levels.

1. Nonexistent: Have not addressed this capability as an organization
2. Initial: Have made some initial attempts at this but are just getting started, more ad hoc and HR driven
3. Defined: Have spent some time building this capability but not fully integrated it yet
4. Managed: Well understood and implemented in the company; management has buy-in
5. Optimized: Have tracked and improved this capability; fully integrated and driven by executive management

Figure 9.2 shows how the maturity levels can be matched against specific strategic leadership development practices and how the value of the

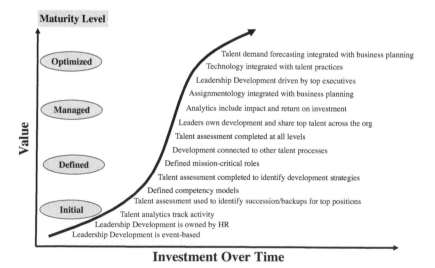

FIGURE 9.2
Strategy-driven leadership development maturity example.

practices can increase over time. The actual practices and shape of the maturity curve may change from organization to organization, but the concept is the same.

We've found that this process visual has been very effective in describing the SDLD journey to a variety of audiences but especially to top executives. Talent Management leaders have used the maturity curve to highlight that this work is an ongoing, continuous improvement initiative and not a proverbial one-and-done." It matches what we have found to be a very effective approach to selling ideas to the C-Suite which involves:

- Clearly describe where the organization is currently on a capability and what the strategic consequences will be if nothing is done to improve the maturity of the capability
- Describe where the organization needs to be because of strategic demands and the benefits of advancing the maturity to this level
- Outline concrete step-by-continuous improvement-steps to get there, probably over several years.

Where is your organization in terms of where it falls along Strategy-Driven Leadership Development maturity curve? Where does it need to be? How are you going to get there?

B. STRATEGY-DRIVEN LEADERSHIP DEVELOPMENT DERAILERS

Your journey to a fully integrated, supported, and effective approach to Strategy-Driven Leadership Development can sometimes become side-tracked or even stalled. To help you avoid that outcome, we have highlighted some typical SDLD "derailers," the impact they can have on the organization, and some remedies on how the derailers can be avoided or addressed. We have used these 12 elements in the form of an assessment to identify opportunities to accelerate the journey.

1. **Not in My House**—Functional managers protect talent that reports to them, see them as "their employees" and do not offer them as

candidates for advancement or development opportunities. They may even get upset if other managers talk to "their people" about opportunities.

The Impact

Talent pools are limited and do not reflect the real bench strength of the organization. Agile, high-potential, and high-professional talent becomes frustrated and leaves the company. Alternatively, their careers may derail due to over-reliance on a single benefactor who is not supporting their career development and advancement.

Remedies

- Build a strong business case for talent from the beginning, driven by strategy. Develop a marketing plan for the business case that targets key stakeholders in all functions/businesses.
- Do not base talent discussions on the input of a single functional manager. The focus should be on having rich talent discussions with input from all available sources. Do not use manager-only performance reviews as input to your talent process.
- Include "Skip-level" meetings in the career development plans for key talent so that executives outside of the function are aware of the talent and can serve as sponsors.
- Include company-wide talent development targets in executive performance and incentive plans. This should be the same measure or objective for all leaders so that "we're all in this together" is reinforced.
- Measure the ratio of High Potential-to-Others in a function's bench strength and challenge the manager if it is low compared to other functions.
- Do not allow leaders in key roles to select their own successors. Set the policy that identifying and selecting replacements for leaders in strategic, pivotal, and scarce roles is not solely the responsibility of the functional manager but that the succession for those roles is owned by a team of executives higher in the organization.
- Assure that building internal and external networks is a key component of Intentional Development Plans for High Professionals,

Adaptable Professionals, Emerging Leaders, and High Potentials. Identify specific opportunities, projects, or assignments that get employees in these talent pools exposed to top executives.

2. **Lists for Lists' Sake**: Completing talent discussions just to have a succession plan or backup list.

The Impact

A we mentioned in an earlier chapter, we have observed that the actual hit rate on backup lists is typically only about 15%, contributing little to improving organization capability and providing a low return for the effort. The true capability of the total organization to achieve strategic targets is not assessed or improved. In addition, High Professional (as opposed to High Potential) talent, critical to maintaining a company's core competencies, is overlooked and may become disengaged. In addition, a large pool of talent that is often "under the radar" but has significant potential to grow is overlooked.

Remedies

- Drive talent discussions from broad organization needs derived from the strategy, not just management succession.
- Complete talent planning based on broad talent pools at all levels, not just the top positions.
- Review and update talent on a regular basis in conjunction with monthly, quarterly, and annual business reviews.
- Build detailed talent development action plans based on the output from and immediately after talent reviews. Assign accountability and track follow-up.
- Establish other measures of the impact of Strategy-Driven Leadership Development (i.e., internal placement rate for strategic, pivotal, and scarce roles, percentage of High Professionals, Adaptable Professionals, Emerging Leaders, and High Potentials with Intentional Development Plans, success of new hires placed in strategy-critical roles, etc.)

3. **Take a Pill**: Development of high potential talent is limited to "take a training class" rather than meaningful assignments or broader-based development.

Impact

Talent is slowly developed or not developed at all. High Professionals, Adaptable Professionals, Emerging Leaders, and High Potentials employees get frustrated by a lack of challenging work and growth opportunities. Training budgets are usually "fickle," so development can be erratic if training is over-emphasized. The return on training investment is not realized. The same people get tapped over and over for key projects.

Remedies

- Curate development experiences. Develop an Assignment Management process that considers talent for all critical assignments—project, process, and business—not just leadership positions. Organizations often add this to their approach to Project Management Offices (PMO).
- Always build an Impact Map before any intervention is selected. Build the line of sight logic.
- Follow the 70–20–10 model of development. Build career development and coaching processes based on the 10 Intentional Leadership Development Essentials.
- Strive to make all training just-in-time and action-based, involving intact project or work teams.
- Measure all training at the impact and behavior level, not just at the reaction and learning level, using the Success Case Evaluation Method.

4. **HR Disconnect**: The talent management process is disconnected from other HR systems such as selection, performance management, compensation, and career development.

Impact

Employees are confused by differing feedback. Users of the HR processes—managers and employees—do not see the benefit in following them. The utility (cost/benefit) of the other processes is not realized. The link between the HR processes and business needs is lost, causing them to be seen as "administrivia."

Remedies

- Build HR processes upon competency-driven technology so there is a common language for talent, no matter the application. Create future-focused competency models based on behaviors of high-performing employees, not the average or typical performer.
- Assign responsibility for all the talent-related processes to one function. Remove transactional processes (benefits, payroll, FMLA, pension) from the talent management function.
- Implement software systems that integrate performance management, compensation, career planning, and talent management.
- Train talent management professionals in behavior-based interviewing and to be facilitators of the talent review process.

5. **Strategy Disconnect**: Talent Management is not driven from strategy and business planning.

Impact

Talent is seen as the responsibility of HR. The true capability of the company to achieve strategic targets is not improved. Critical budgeting and resourcing for talent is missed. Time devoted to talent reviews and talent development is viewed as a waste of time.

Remedies

- Build clear links in the business planning process to the talent management process. Match the talent management cycle to the business planning and budgeting cycle.
- Add talent management measures to overall company scorecards. Analyze sensitivities in the measures regularly.
- Ask talent-related questions at key junctures in the planning process—"Do we have the organization capability to achieve these goals?" Analyze the results of talent reviews to answer strategic talent questions.
- Implement Employee Performance Management software that links goal alignment to talent management.

6. **Nothing Gained:** A lack of or inconsistent follow-up on organization development action items from talent reviews.

Impact

The long-term capability of the organization is not improved. Initial momentum for the process is eventually lost. Employees see a varying level of support for the process and wonder what's happening. Key talent as well as managers may become frustrated because the only hear words and see no action.

Remedies

- Add talent management objectives to leadership performance plans and link managers' compensation to the performance on the objectives.
- Add a review of progress on talent review Action Items to regular business meetings.
- Provide consulting or coaching support to managers with significant talent issues.
- Build in talent discussions to regular meetings such as strategy updates and planning sessions.

7. **It's Not My Job**: Managers depend solely on HR or outside coaches for development.

Impact

Talent management that is not owned and implemented by management goes nowhere. Progress lags. Coaching is focused on the wrong players, providing little return on the investment. The talent dialogue and process is not continuously improved. Talent issues, both positive and negative, are not addressed quickly or effectively.

Remedies

- Build coaching, feedback, and Intentional Development concepts into all training for leaders/managers/supervisors. Make it a competency requirement in their performance plan.
- Select managers based on their ability to develop others (a hallmark of learning agile leaders).
- Remove non-developing (low performing, low potential) managers from strategy-critical positions where talent development and performance is crucial.

- Starting with the business case, build management responsibility for talent into the process. Don't proceed unless there is a clear case and commitment. (See Frame It.)
- Plan for and provide just-in-time development for managers in the talent process (usually immediately following a talent review).

8. **Information Black Hole**: Not having a robust database to track talent information and progress.

Impact

Talent data stays as data and does not become "talent intelligence." There is little understanding of whether the talent process is under control or not, so it's unclear when corrective action is needed. Problems (turnover, productivity, no backup) are only recognized when it's too late to address them. The process is paper-intensive, and managers and employees see it as a burden. The benefit of talent management initiatives is questioned.

Remedies

- Define the process first but back it up with a robust database and measurements. Build the process from business requirements, not HR requirements alone.
- Define the data needs in advance and work with IT to understand requirements. Do a pilot in a smaller area to confirm the system needs.
- Measure the productivity of the process with and without effective data (time spent by key executives) to build a case.

9. **Just the Facts**: A clearly defined and accepted business case for Talent Management does not exist.

Impact

The true capability of the organization's talent to achieve business goals is not identified. Resources are not allocated effectively. The process does not move beyond a very low level of maturity and loses support over time.

Remedies

- Build the business case before you do anything else. Use a cross-functional team to define the effort, ideally not led by HR. Talk to key stakeholders and identify talent issues linked to business needs and drivers. Complete a "What Works" analysis to identify talent practices that make a difference and combine them in a fashion for your organization that is innovative.
- Define and measure "Talent Management" early in the process. Review the measures with key stakeholders and report them during business reviews.
- Don't just present the findings but use key influencers in your organization to carry the message. Be certain to use all levels of employees as influencers and not just senior level people.

10. **Only at the Top**: Having rich talent discussions only for top-level positions.

Impact

Future talent gaps are not identified and become a barrier to growth. The capacity to complete key projects or to grow does not improve. No benefit is realized from knowing what talent is available for resource-constrained projects. The total organization capability is not assessed. Turnover may occur at middle levels because employees do not see an opportunity to grow. There is an overemphasis on buying talent (hiring) rather than creating it from within.

Remedies

- Start at the top but lay immediate plans to cascade talent reviews and action planning down through the organization.
- Plan for and create a process to assess talent *pools* at all levels—executive, senior manager, key manager, and an early watch list.
- Implement manager and employee self-service software that can empower lower level supervision to contribute to the talent process.

11. **Resurrection Is Much Harder Than Birth**: Spending too much time and resources on low learning agile, low potential employees, trying to fix them.

Impact

Talent improvement is slowed. A return on investment in Human Capital is not realized because the investment is going to those with a very low chance of improvement. High potential talent becomes frustrated and leaves.

Remedies

- Consider the talent review process as identifying different development strategies, not creating backup lists.
- Assure that marginal employees are identified and that quick, "short fuse" action plans are established. Focus on pivotal roles and functions first.
- Place full responsibility for change on the blocked employee's shoulders with clear and quick consequences if change does not occur. Do not "take the monkey."
- Build consequences for managers that do not address performance issues quickly.

12. **Living in the Past**: Not advancing from a focus on the past to a focus on the future as the talent process matures.

Impact

Talent gaps are missed and hinder growth. Too much time is spent on reacting to talent problems rather than creating opportunities and continuously improving organization capability. The business case for talent management is not validated.

Remedies

- Implement an information system that allows you to model talent over time—"what if"—and to track progress over previous years.
- Gather operations, sales, and business planning data as input to the talent process.
- Include a discussion of talent implications to business plans.
- Measure the impact of talent progress on business results. Look for talent sensitivities.

10

Conclusion

A LEADERSHIP DEVELOPMENT FAIRY TALE . . . THE NEXT YEAR

It had been a year since the new CEO came onboard and today, he was meeting with his team to discuss whether their annual leadership event would come to pass. It had been a good year for leadership development, and the boss wanted to use their full agenda to talk about his team's point of view.

He recalled to them how they ran these programs in the past. They were worthy events, held at exclusive resorts over a multiday period. They rewarded people for their hard work and diligence, provided some motivational and one-shot ideas but, he noted, they yielded little in how they company used the learnings from the program to meet and grow the strategic objectives of the business.

He wanted to know how they perceived the shift to Strategy-Driven Leadership Development. Had this new effort changed their perspective on talent management and more importantly, how had it impacted the business? He asked his team for their observations about the past year. Here is what they shared.

His COO spoke first. He told his colleagues that he was very pleased with how well his leadership team had embraced Intentional Leadership Development and that he could see how the leadership within the business's operations was improving. When the CEO pressed him for some specifics, he shared that for the first time he was seeing mid-level managers take greater responsibility for driving business actions. In the past, they had been "managing" their teams, making certain that the work in their departments got done. Now, however, they were generating new ideas, engaging their teams in decision making and quality improvement activities, and had created a new sense of pride in how work got done.

He also observed how "development had rolled downhill." As his leaders focused on their own skill enhancement, they had, in turn, challenged their team members to be more intentional about their development by building Impact Maps, Learning Maps, and Development Action Plans and by doing regular reflection and development tracking.

Next up was his CTO, who immediately got a laugh when he mentioned about how much more sociable his geek IT leadership group had become as a result of the leadership program. Seriously, he went onto say, that while the leadership skills his team gained were valuable, what proved most important to him was that his team learned the importance of building strong organizational networks from the talent reviews and actually formed new relationships with other leaders in the organization due to the leadership cohort groups. He shared that while people laugh about his team's lack of interpersonal skills, he found that the program not only helped them improve their skills but allowed them to work on improving how they influenced the organization. They learned that there was more than one way to build competence and not to just focus on perceived weaknesses.

The CMO was more modest in her description of the program's results for her team. While leadership skills are always important, she notes, business success for her hinged on getting her people focused on their customers and using technology to drive new and existing business relationships. Leadership was important but she felt that her team's leadership skills were satisfactory and that the focus on building leadership skills did not bring much value to the team. She recognized however that given the fluidity with which salespeople may jump from job to job, having strong leaders in this group could only improve people's desire to stay with the firm. She felt her team could do a better job of creating Impact Maps to more clearly establish the line of sight between their skills and key business outcomes and to build development into work and not view it as something extra they needed to do. They also had to work together to improve their overall ability to assess Growth Potential and Performance Effectiveness down through the marketing function.

The CHRO reported on the analytics derived from the systematic talent reviews. He pointed out that having tools to understand how the organization aligned around talent provided his team with an overarching strategy that, he believed, made his team more valuable to their business partners. His colleagues nodded in agreement. Most importantly, he commented that his team was no longer just reacting to employee relations issues but instead were proactively focusing on curating development opportunities for aspiring, high-potential leaders and improving the overall health of the organization's leadership pipeline at all levels. Being able to provide his colleagues with analytics such as internal placement rates, bench strength, talent pipeline capability. and retention projections clearly brought his HR group into being a strategic business partner. The focus on integrating competencies into all of the company's talent processes also created an alignment within HR that had not existed before. He was very pleased to hear leaders across the company referring to the competencies, all the way up to the CEO.

The CFO earned her stripes and success by carefully managing every dollar that went through the organization, and she was initially not in favor of the program. She didn't think that having a targeted leadership program would yield the results that would match the costs, but, she admitted, she was wrong. She shared that the first hint of success for her was when one of her own high potential direct reports, who she knew was thinking about leaving, came to her and shared that he was excited about his work with an Intentional Development cohort and that he appreciated the company investing in him. Her insight from that one experience helped her to recognize an immediate direct savings from not having to go through a rehire for that position but also for the indirect savings of the cost of the headaches and aggravations that she would have faced if she had to bring on and train a new person. Now that she saw that kind of direct benefit, she was herself open to look at the other strategic values the program could provide.

The CEO Went Last

He started by observing the improved leadership skills he saw in each of the team members sitting at the table that day. Each one of them made a commitment to his/her own leadership development, and he could easily point out the gains they had made for themselves. He also liked that as a result of regular feedback and reflection on their development, the team itself was more honest and sharing of concerns in their meetings . . . that kind of gain and improvement was invaluable. He even went onto point out that his own intentional skill improvement, which was to dial down his overuse of "getting into the weeds" helped him to focus on bigger strategic and marketing activities, which was a better use of his expertise and that his letting the team run their businesses had helped the business grow this past year. Everyone agreed with a respectful smile. Finally, he pointed out that while the costs of time and money had been significant for this project, it was now theirs. They owned the leadership connections to their business strategy. They owned the competency models that would drive business success, and they understood how to change those models as business demands required. They owned the process of how to intentionally coach their employees in their leadership development. They owned the regular and systematic review of talent at all levels, matched against the strategic demands of the business. And they owned the cultural shifts that were happening as a result of employees seeing the firm investing in their future.

He even surprised them by announcing that they were going to have another off-site leadership program this year. It would be more modest than in the past. Held at a local resort, it would just be one day, but it would be a day to celebrate their accomplishments. Instead of bringing in famous speakers from the outside, they would spend the morning hearing from leaders throughout the firm who had partnered up with their colleagues to share their new projects, ideas, and accomplishments. As in previous years, they would still have an afternoon of fun, which was well earned by all.

Even though we opened and are closing this book with fairy tales (both of which have elements of the challenges and results we've seen from a variety of our clients,) fairy tales do not present the challenges facing our businesses today. The challenges facing leadership and leadership development have become more complex. Companies globally are taking steps to disrupt their competitive landscape which creates a compelling case to reshape their current leaders and the process they use to identify and develop new leaders.

We followed our opening fairy tale with a description of the journey that we have taken to view leadership development from a more strategic perspective. Strategy-Driven Leadership Development is built on the foundation that all talent practices should be derived from strategy, not the latest fad or time-worn assumptions. From this basic strategic foundation, SDLD evolved from a number of evidence-based beliefs and practices.

First, we know that effective leaders are not just born and that key leadership skills are not innate gifts from on high. All critical leadership skills can be effectively developed if an organization views its talent from a growth mindset and follows 10 essentials of Intentional Leadership Development.

Second, organizations do not have unlimited resources to identify and develop talent. Employees do not all perform the same and do not have the same potential or aspiration to grow. Organizations can best assure a return on their talent investment by effectively assessing talent and matching different development strategies to similarly abled pools of talent. The investment can be further enhanced by not focusing on any and all positions but by investing the most in the development of leaders in strategy-critical roles and their potential replacements.

Third, we know that most past approaches to leadership development have not worked. Event-based or one-and-done leadership workshops are not effective in building complex leadership competencies needed in today's business environment. Enhanced and new leadership skills come from navigating a variety of challenges and intentionally learning from them. Development must be immersive. It must be built into our everyday work, not bolted on as something extra that leaders get to when they have the time.

Fourth, learning agility is the new growth hormone. Some employees go through challenging, developmental experiences but do not learn new competencies. We can enhance leadership development by understanding the components of learning agility and assisting learners in building new strategies for learning from experience. We can match development strategies to the talent and not think that one approach can work for all types of talent

Fifth, development is most effective when it is *intentional*. That intentionality comes from a focused four-step process—Frame It, See It, Own It, and Connect It. The Intentional Leadership Development approach can be used in one-on-one coaching, in development cohorts, and in enhancing the effectiveness of existing structured leadership development programs. Intentional Development best occurs in cohorts of learners, involves using simple mantras that trigger and focus learning, leverages constant reflection, and requires a regular cadence of development—development that occurs on a daily or weekly pace.

Sixth, effective leadership is about employing the right behaviors at the right time. The "right behaviors" are best described as competencies that are directly linked to an organization's strategy. The behaviors don't come from a list in the most recent best-seller, by copying other company's competency models, or from joining the cult of personality traits.

Seventh, feedback and reflection are essential to development and can enhance what is learned from challenging experiences. But feedback must be requested by a learner and not imposed. Feedback can come from a variety of sources, but it should be expressed in behavioral terms (competencies). Anyone in a leadership role must be an intentional coach. Leaders must be effective at creating the environment where feedback is welcomed and it is provided in a timely and effective manner. The environment should encourage learning from successes as well as failures.

Finally, building an optimized approach to Strategy-Driven Leadership Development is a journey that takes time and continuous improvement. The journey starts by plotting out where your organization is currently and building a clear path and business case to a future target. There can be a number of threats that can derail the process along the way, but remedies exist for getting your talent train back on track.

We close the book with our second fairy tale, which imagined the gains made in key business areas and how those gains produced improved business outcome for the organization. The connections our clients built over time from these varied and evidence-based sources eventually fell together, for them, as Strategy-Driven Leadership Development and Intentional Development. As a result of following the talent evidence, we've helped numerous businesses to significantly alter many of their assumptions about learning, the language they employed around developing leaders, and the steps they could take to develop leaders who drive business success.

The journey for Strategy-Driven Leadership Development journey is not at an end. It is actually just beginning. We expect that new and expanded research will continue to evolve the model and that enhanced ways to build Intentional Development will allow our clients and readers to continually update our development model and tools. We can't wait to see what comes next.

References

Avedon, M. J., & Scholes, G. (2010). Building competitive advantage through integrated talent management. In R. F. Silzer & B. E. Dowell (Eds.), *Strategy-driven talent management: A leadership imperative*. San Francisco, CA: Jossey Bass.

Balkundi, P., & Harrison, D. A. (2006). Ties, leaders, and time in teams: Strong inference about network structure's effects on team viability and performance. *Academy of Management Journal, 49*, 49–68.

Barends, E., Rousseau, D. M., & Briner, R. B. (2014). *Evidence-based management: The basic principles*. Amsterdam, The Netherlands: Center for Evidence-Based Management. Retrieved from www.cebma.org/wp-content/uploads/Evidence-Based-Practice-The-Basic-Principles-vs-Dec-2015.pdf

Bass, B., Avolio, B., Jung, D., & Berson, Y. (2003). Predicting unit performance by assessing transformational and transactional leadership. *Journal of Applied Psychology, 88*(2), 207–218.

Becker, B. E., Huselid, M. A., & Beatty, R. W. (2009). *The differentiated workforce*. Boston, MA: Harvard Business Press.

Bishop, C. H., Jr. (2001). *Making change happen one person at a time: Assessing change capacity within your organization*. New York, NY: AMACOM.

Bloom, N., Dorgan, S., Dowdy, J., Van Reneen, J., & Rippin, T. (2005). *Management practices across firms and nations*. London: Centre for Economic Performance and McKinsey and Company.

Boatman, J., & Wellins, R. S. (2011). *Global leadership forecast*. Pittsburgh, PA: Development Dimensions International.

Bontis, N., Hardy, C., & Mattox, J. R. (2011). *Diagnosing key drivers of job impact and business results attributable to training at the Defense Acquisition University*. Fort Belvoir, VA: Defense Acquisition University.

Boudreau, J. W. (2010). *Retooling hr. Using proven business tools to make better decisions about talent*. Boston, MA: Harvard Business Press.

Boudreau, J. W., & Ramstad, P. (2007). *Beyond HR: The new science of human capital*. Boston, MA: Harvard Business Press.

Boyce, A. S., Nieminen, L. R. G., Gillespie, M. A., Ryan, A. M., & Denison, D. R. (2015). Which comes first, organizational culture or performance? A longitudinal study of causal priority with automobile dealerships. *Journal of Organizational Behavior, 36*(3), 339–359. doi:10.1002/job.1985

Brinkerhoff, R. (2006). *Telling training's story: Evaluation made simple, credible and effective*. San Francisco, CA: Barrett-Koehler.

Brousseau, K. R., Driver, M. J., Hourihan, G., & Larsson, R. (2006, February). The seasoned executive's decision-making style. *Harvard Business Review*. Retrieved from https://hbr.org/2006/02/the-seasoned-executives-decision-making-style

Buckingham, M., & Clifton, D. O. (2001). *Now, discover your strengths*. New York, NY: Simon & Schuster Inc.

Burke, W. W., & Noumair, D. A. (2002). The role of personality assessment in organization development. In J. Waclawski & A. H. Church (Eds.), *Organization development: A data-driven approach to organizational change* (pp. 55–77). San Francisco, CA: Jossey–Bass.

Byford, M., Watkins, M., & Triantogiannis, L. (2017, May–June). Onboarding isn't enough. *Harvard Business Review*. Retrieved from https://hbr.org/2017/05/onboarding-isnt-enough

Byham, W. C., Smith, A. B., & Pease, M. J. (2015). *Grow your own leaders: How to identify, develop and retain leadership talent.* Upper Saddle River, NJ: Prentice-Hall Inc.

Cappelli, P. (2008). *Talent on demand: Managing talent in an age of uncertainty.* Boston, MA: Harvard Business Press.

Cavanaugh, C., & Zelin, A. (2015). *Want more effective managers? Learning agility may be the key.* Bowling Green, OH: Society for Industrial and Organizational Psychology. Retrieved from www.siop.org/WhitePapers/LearningAgilityFINAL.pdf

CEB. (2013). *Breakthrough performance in the new work environment.* Executive Guidance for 2013. Retrieved from www.cebglobal.com/content/dam/cebglobal/us/EN/top-insights/executive-guidance/pdfs/eg2013ann-breakthrough-performance-in-the-new-work-environment.pdf

CEB. (2014). *Succession management for the new work environment: From pipeline to portfolio management.* Retrieved from www.assessmentanalytics.com/wp-content/uploads/2014/12/CEB-Succession-Planning-Presentation-Deck.pdf

Charan, R., Drotter, S., & Noel, J. (2001). *The leadership pipeline: How to build the leadership-powered company.* San Francisco, CA: Jossey-Bass Inc.

Church, A. H., & Waclawski, J. (2010). Take the Pepsi challenge: Talent development at Pepsico. In R. F. Silzer & B. E. Dowell (Eds.), *Strategy-driven talent management: A leadership imperative.* San Francisco, CA: Jossey Bass.

Citrin, R. S., & Weiss, A. (2016). *The resilience advantage: Stop managing stress and find your resilience.* New York, NY: Business Expert Press.

Conger, J. (2010). Developing leadership talent: Delivering on the promise of structured programs. In R. F. Silzer & B. E. Dowell (Eds.), *Strategy-driven talent management: A leadership imperative.* San Francisco, CA: Jossey Bass.

Connor, J. (2011). Deepening the talent pool through learning agility. *People Management*, pp. 40–43. Retrieved from www.peoplemanagement.co.uk/pm/articles/2011/11/deepening-the-talent-pool-through-learning-agility.htm

Corporate Leadership Council. (2005). *Realizing the full potential of rising talent (Volume I): A quantitative analysis of the identification and development of high-potential employees.* Washington, DC: Corporate Executive Board.

Cross, R., & Prusak, L. (2002). The people who make organizations go—or stop. *Harvard Business Review*, *80*(6), 104–112. Retrieved from https://hbr.org/2002/06/the-people-who-make-organizations-go-or-stop

Curtis, B., Hefley, W., & Miller, S. (2001). *People capability maturity model.* Version 2.0, CMU/SEI-2001-MM-01. Pittsburgh, PA: Carnegie Mellon Software Engineering Institute.

Dai, G., De Meuse, K. P., & Tang, K. Y. (2013). The role of learning agility in executive career success: The results of two field studies. *Journal of Managerial Issues*, *25*(2), 108–131.

Davachi, L., Kiefer, T., Rock, D., & Rock, L. (2013). Learning that lasts through AGES. In D. Rock & A. H. Ringleb (Eds.), *Handbook of neuroleadership* (pp. 463–480). Lexington, KY: Neuroleadership Institute.

De Meuse, K. P. (2017). Learning agility: Its evolution as a psychological construct and its empirical relationship to leader success. *Consulting Psychology Journal*, *69*(4), 267–295.

De Meuse, K. P., Dai, G., & Hallenback, G. S. (2010). Learning agility: A construct whose time has come. *Consulting Psychology Journal*, *62*(2), 119–130.

Denison, D., & Hallagan, R. (2017, April–May). *The board perspective: Does corporate culture matter*. Retrieved from www.denisonconsulting.com/transform/article-corporate-culture-matter/

DeRue, D., & Wellman, N. (2009). Developing leaders via experience: The role of challenge, learning orientation and feedback availability. *Journal of Applied Psychology, 94*(4), 859–875.

Di Stefano, G., Gino, F., Pisano, G., & Staats, D. (2016). *Making experience count: The role of reflection in individual learning* (Harvard Business School NOM Working Paper No. 14–093). Retrieved from https://papers.ssrn.com/sol3/papers.cfm?abstract_id=2414478##

Dixon, P., Rock, D., & Ochsner, K. (2013). Turn the 360 around. In D. Rock & A. II. Ringleb (Eds.), *Handbook of neuroleadership* (pp. 447–462). Lexington, KY: Neuroleadership Institute.

Dries, N., Vantilborgh, T., & Pepermans, R. (2012). The role of learning agility and career variety in the identification and development of high potential employees. *Personnel Review, 41*(3), 340–358.

Dweck, C. (2016). *Mindset: The new psychology of success*. New York, NY: Ballantine Books.

Eichinger, R. W., Lombardo, M. M., Stiber, A., & Orr, J. E. (2011). *Paths to improvement: Navigating your way to success*. Minneapolis, MN: Korn/Ferry International.

Enderes, K., & Deruntz, M. (2018). *Seven top findings for moving from managing performance to enabling performance in the flow of work*. Deloitte Development LLC. Retrieved from login.bersin.com/uploadedFiles/092418_RF_HIPMTopFindings_KE_Final.pdf

Fernandez-Araoz, C. (2014, June). 21st century talent spotting. *Harvard Business Review*. Retrieved from https://hbr.org/2014/06/21st-century-talent-spotting

Feser, C., Nielson, N., & Rennie, M. (2017, August). What's missing in leadership development. *McKinsey Quarterly*. Seattle, WA: McKinsey & Comp.

Fitz-Enz, J., & Mattox, J. (2014). *Predictive analytics for human resources*. Hoboken, NJ: Wiley.

Flaum, J. P. (2009). *When it comes to business leadership, nice guys finish first*. Green Peak Partners. Retrieved from https://greenpeakpartners.com/wp-content/uploads/2018/09/Green-Peak_Cornell-University-Study_What-predicts-success.pdf

Freedman, A. M. (2011, January). *Meeting the challenges of moving through the leadership pipeline*. Presentation at the Society of Consulting Psychology. Retrieved from http://kaplandevries.com/images/uploads/MeetingPipelineChallenge_SCP2011mwc.pdf

Fuller, T. (2019). *Differentiating strengths*. Compio. Retrieved from https://compio.net/leveraging-strengths/differentiating-strengths/

Gandossy, R., & Effron, M. (2003). *Leading the way: Three truths from the top companies for leaders*. Hoboken, NJ: John Wiley and Sons, Inc.

Garvin, D. (2013, November–December). How google sold its engineers on management. *Harvard Business Review*. Retrieved from https://hbr.org/2013/12/how-google-sold-its-engineers-on-management.-

Gitsham, M. (2009). *Developing the global leader of tomorrow*. Ashridge, UK: Ashridge Business School. Retrieved from ashridge.org.uk

Gollwitzer, P. M. (1999). Implementation intentions: Strong effects of simple plans. *American Psychologist, 54*(7), 493–503.

Gurdjian, P., Halbeisen, T., & Lane, K. (2014, January). Why leadership-development programs fail. *McKinsey Quarterly*. Seattle, WA. Retrieved from www.mckinsey.com/featured-insights/leadership/why-leadership-development-programs-fail

Hagemann, B., & Mattone, J. (2011). *2011/2012 trends in executive development: A benchmark report*. Oklahoma City, OK: Executive Development Associates (EDA) Inc. and Pearson Education Inc.

Hansen, G. S., & Wernerfelt, B. (1989). Determinants of firm performance: The relative importance of economic and organizational factors. *Strategic Management Journal, 10*(September–October), 399–411.

Huselid, M. (1995). The impact of human resource management practices on turnover, productivity, and corporate financial performance. *The Academy of Management Journal, 38*(3), 635–872.

Ingvar, D. H. (1985). Memory of the future: An essay on the temporal organization of conscious awareness. *Human Neurobiology, 4*(3), 127–136. Retrieved from www.ncbi.nlm.nih.gov/pubmed/3905726

Jacobs, K. (2015, May). Coaching and feedback still rare in performance management. *HR Magazine*. London, UK. Retrieved from http://www.hrmagazine.co.uk/article-details/coaching-and-feedback-still-rare-in-performance-management

Jensen, E. (2005). *Teaching with the brain in mind* (2nd ed.). Alexandria, VA: Association for Supervision and Curriculum Development.

Kesler, G. C. (2002). Why leadership bench never gets deeper: Ten insights about executive talent development. *HR Planning Society Journal, 25*(1). Retrieved from http://citeseerx.ist.psu.edu/viewdoc/download?doi=10.1.1.461.6090&rep=rep1&type=pdf

Knowles, M. (1984). *The adult learner: A neglected species* (3rd ed.). Houston, TX: Gulf Publishing.

Korn Ferry. (2014). *Korn Ferry leadership architect research guide and technical manual*. Korn Ferry. Retrieved from www.kornferry.com

Krackhardt, D., & Hanson, J. R. (1993, July–August). Informal networks: The company behind the chart. *Harvard Business Review*. Retrieved from https://hbr.org/1993/07/informal-networks-the-company-behind-the-chart

Lamoureux, K., Campbell, M., & Smith, R. (2009, April). *High-impact succession management: Executive summary*. Bersin & Associates and Center for Creative Leadership Industry Study. Retrieved from https://files.eric.ed.gov/fulltext/ED507599.pdf

Ledford, G. E., & Schneider, B. (2018). *Performance feedback culture drives business impact*. Institute for Corporate Productivity and the Center for Effective Organizations. Retrieved from https://ceo.usc.edu/files/2018/07/Performance-Feedback-Culture-Drives-Business-Performance-i4cp-CEO-002.pdf

Loew, L. (2015). *State of leadership development 2015: The time to act is now*. Brandon Hall Group. Retrieved from www.ddiworld.com/DDI/media/trend-research/state-of-leadership-development_tr_brandon-hall.pdf

Loew, L., & Garr, S. (2011). *High-impact leadership development: Best practices for building 21st-century leaders* (pp. 113–116). Oakland, CA: Bersin & Associates.

Lombardo, M. M., & Eichinger, R. W. (2000). High potentials as high learners. *Human Resource Management, 39*(4), 321–329.

Lombardo, M. M., & Eichinger, R. W. (2011). *The leadership machine: Architecture to develop leaders for any future*. Minneapolis, MN: Lominger International.

Majdan, K., & Wasowski, M. (2017, April). *We sat down with Microsoft's CEO to discuss the past, present and future of the company*. Business Insider Poland. Retrieved from www.businessinsider.com/satya-nadella-microsoft-ceo-qa-2017-4

Martin, K., & Armitage, A. (2016). *Talent risk management*. Institute for Corporate Productivity (i4cp). Retrieved from www.i4cp.com/company/downloads

McCall, M. W., & Hollenbeck, G. P. (2002). *The lessons of international experience: Developing global executives*. Boston, MA: Harvard Business School Publishing.

McCall, M. W., Lombardo, M., & Morrison, A. (1988). *The lessons of experience: How successful executive develop on the job.* New York, NY: The Free Press.

McCauley, C. D., Ruderman, M. N., Ohlott, P. J., & Morrow, J. E. (1994). Assessing the developmental components of managerial jobs. *Journal of Applied Psychology, 79*(4), 544–560.

Meister, J., & Willyerd, K. (2010, May). Mentoring millennials. *Harvard Business Review.* Retrieved from https://hbr.org/2010/05/mentoring-millennials

Mitchel, C., Ray, R. L., & van Ark, B. (2017). *The conference board CEO challenge 2017: Leading through risk, disruption and transformation.* New York, NY: The Conference Board.

Mooney, T., & Brinkerhoff, R. (2008). *Courageous training: Bold actions for business results.* San Francisco, CA: Berret Koehler.

Moules, J., & Nilsson, P. (2017, August). What employers want from MBA graduates— and what they don't. *Financial Times.* Retrieved from https://www.ft.com/content/64b19e8e-aaa5-11e8-89a1-e5dc165fa619

Mueller-Hanson, R. A., White, S. S., Dorsey, D. W., & Pulakos, E. D. (2005). *Training adaptable leaders: Lessons from research and practice* (Research Report 1844). Minneapolis, MN: Personnel Decisions Research Institutes.

New Talent Management Network. (2015). *Potential: Who's doing what to identify their best.* The Talent Strategy Group. Retrieved from www.talentstrategygroup.com/application/third_party/ckfinder/userfiles/files/NTMN%20Potential%20Study%202015.pdf

Okpara, A., & Edwin, A. (2015). Self-awareness and organizational performance in the Nigerian banking sector. *European Journal of Research and Reflection in Management Science, 3*(1).

O'Leonard, K., & Loew, L. (2012). *Leadership development factbook 2012: Benchmarks and trends in U.S. leadership development.* Oakland, CA: Bersin & Associates.

Paul, A. M. (2004). *The cult of personality testing: How personality tests are leading us to miseducate our children, mismanage our companies and misunderstand ourselves.* New York, NY: Free Press.

Peshawaria, R. (2011, November). The great training robbery: Why the $60 billion investment in leadership development is not working. *Forbes.com.* Retrieved from https://www.forbes.com/sites/rajeevpeshawaria/2011/11/01/the-great-training-robbery-2/#4a375dd61deb

Peterson, D. (2006). People are complex and the world is messy: A behavior-based approach to coaching. In D. Stober & A. Grant (Eds.), *Evidence-based coaching handbook: Putting best practices to work for your clients.* Hoboken, NJ: Wiley.

Robinson, G. S., & Wick, C. W. (1992). Executive development that makes a business difference. *Human Resource Planning, 15*(1), 63–76.

Rock, D. (2009, August). *Managing with the brain in mind. Strategy+Business,* (56). Reprint number 09206. Retrieved from www.strategy-business.com/article/09306?gko=5df7f

Rock, D., & Ringleb, Al. (2013). *The handbook of neuroleadership.* Lexington, KY: The NeuroLeadership Institute.

Russell, C. J. (2001). A longitudinal study of top-level executive performance. *Journal of Applied Psychology, 86*(4), 560–573.

SAP SuccessFactors and Workforce Intelligence Institute. (2006). *How smart human capital management drives financial performance.* Retrieved from www.successfactors.com/content/ssf-site/en/resources/knowledge-hub/educational-articles/how-smart-human-capital-management-drives-financial-performance.html

Sarkar, A., Fienberg, D. E., & Krackhardt, D. (2010). Predicting profitability using advice branch bank networks. *Statistical Methodology, 7,* 429–444.

Sarros, J., Gray, J., & Densten, I. L. (2002). Leadership and its impact on organizational culture. *International Journal of Business Studies, 10*(2), 1–25.

Schein, E. (1996). Culture: The missing concept in organization studies. *Administrative Science Quarterly, 41*(2), 229–235.

Schmidt, F. L., & Hunter, J. E. (1998). The validity and utility of selection methods in personnel psychology: Practical and theoretical implications of 85 years of research findings. *Psychological Bulletin, 124*(2), 262–274.

Scullen, S. E., Mount, M. K., & Goff, M. (2000). Understanding the latent structure of job performance ratings. *Journal of Applied Psychology, 85*(6), 956–970.

Silzer, R., & Church, A. (2009). The pearls and perils of identifying potential. *Industrial and Organizational Psychology, 2*(4), 377–412.

Sinar, E., Wellins, R. S., Ray, R., Abel, A. L., & Neal, S. (2014). *Global leadership forecast 2014|2015 Ready now leaders: Meeting tomorrow's business challenges.* New York, NY: The Conference Board; Pittsburgh, PA: Development Dimensions International.

Sloan, N., Agarwal, D., Garr, S., & Pataskia, K. (2017). *Performance management: Playing a winning hand.* 2017 Deloitte Global Human Capital Trends, p. 65. Retrieved from https://www2.deloitte.com/insights/us/en/focus/human-capital-trends/2017/redesigning-performance-management.html

Swart, T., Chisholm, K., & Brown, P. (2015). *Neuroscience for leadership: Harnessing the brain gain advantage.* London, England: Palgrave Macmillan.

Swisher, V. (2016*). Becoming an agile leader.* Minneapolis, MN: Korn Ferry International.

Talent Management Network. (2015). *Potential: Who's doing what to identify their best?* New York, NY: New Talent Management Network. Retrieved from https://237jzd2nbeeb3ocdpdcjau97-wpengine.netdna-ssl.com/wp-content/uploads/2017/01/NTMN20Potential20Study202015.pdf

Thornton, G. C., & Gibbons, A. M. (2009). Validity of assessment centers for personnel selection. *Human Resource Management Review, 19*, 169–187.

Training Industry Report. (2016, November—December). *Training Magazine.* Retrieved from https://trainingmag.com/sites/default/files/images/Training_Industry_Report_2016.pdf

Ulrich, D., & Lake, D. (1990). *Organizational capability: Competing from the inside out.* New York, NY: John Wiley and Sons.

Watson Wyatt. (2008/2009). *WorkUSA Report, Driving business results through continuous engagement,* released February 10, 2009. Retrieved from https://www.hr.com/en/communities/watson-wyatts-workusa-survey-identifies-steps-to-k_fr1elg39.html

Welch, J. (2001). *Jack: Straight from the gut.* New York, NY: Warner Business Books.

Wentworth, D., & Loew, L. (2013). *Leadership: The state of development programs.* Brandon Hall Group. Retrieved from www.skillsoft.com/assets/research/research_bhg_leadership_development_programs.pdf

Wexley, K., Greenawalt, J., Alexander, A., & Couch, M. (1980). Attitudinal congruence and similarity as related to interpersonal evaluations in manager-subordinate dyads. *Academy of Management Journal, 23*(2), 320–330.

Whitehurst, J. (2015, June–July). Be a leader who can admit mistakes. *Harvard Business Review.* Retrieved from https://hbr.org/2015/06/be-a-leader-who-can-admit-mistakes

Yip, J., & Wilson, M. (2010). Learning from experience. In *The center for creative leadership development handbook* (3rd ed.). San Francisco, CA: John Wiley and Sons.

Zes, D., & Landis, D. (2013). *A better return on self-awareness* (Report). Los Angeles, CA: Korn Ferry Institute.

Index

Note: Page numbers in italics indicate figures and in bold indicate tables on the corresponding pages.

A

accountability, 57
adult learning, concepts in, 69–70
adventure, learning as, 70
aligning and integrating all talent
    processes, 29–30, *30*
aspiration, 42–43
assessment centers, 47
assessment of organization and leadership
    capability, *35*, 35–36
  best tool or method for, **45**, 45–47
  development strategies matrix or nine
    box on steroids for, 58–60, **58–60**
  development strategies roles, *61*, 61–66,
    62, *63*, **65–66**
  looking for growth potential, 37–43
  for performance effectiveness over time,
    43–44
  strategic talent priorities and, 63–64
  Talent Reviews for, 48–57
"Assignmentology," 121

B

Beatty, R. W., 61
Becker, B. E., 19, 61
*Beyond HR: The New Science of Human
    Capital*, 18
Bishop, C., 49
Blackrock, 88
blended learning, 67
blind spots, 130
Boudreau, J., 18, 19, 21
Brache, A., 40
Brinkerhoff, R., 13–15, 91, 141
Buford, J., 34

C

calibration, 53–54
capability in growth potential, 42

career derailment, 129–130
career development vs. development for a
    position, 120
career planning, 136
career transitions, *119*, 119–120, 129
CEB, 44, 86
Center for Creative Leadership, 42, 74,
    79, 119
Center for Effective Organizations, 46
Centre for Economic Performance, 8
change agility, 41
Citrin, R. S., 5, 13, 19, 22, 70
Clifton, D., 19
coaching, definition of, 124; *see also*
    Intentional Leadership Coaching
cohorts, 70, 85–86, 91, 94, 107, 113, 126,
    141–142, 168, 169, 171
competencies, *see* strategy-driven
    competencies
competency flow, 120–121
competitive advantage, 19
Connect it, leveraging new skills through,
    118, 118–122, *119*, 136
Corporate Leadership Council (CLC), 36
Couch, M. A., 13, 18, 19, 22, 40, 48
*Courageous Training: Bold Actions for
    Business Results*, 14
critical competencies, focusing on, 73–74, *74*
Cross, R., 49
curating experiences, 121–122

D

data analytics, 56–57
  using development strategies matrix,
    143–151, *144–150*
derailers, Strategy-Driven Leadership
    Development (SDLD), 157–165
DeRue, D., 138
developmental jobs, 42
development pipeline, 71

development strategies matrix, 58–60, **58–60**
    talent analytics using, 143–151, *144–150*
Differentiated Workforce: Transforming Talent into Strategic Impact, the, 18–19
differentiating strengths, 73–74, *74*
    development targets, 91, 99–101, 158
diversity, 136
Dries, N., 39
Drucker, P., 90
Dweck, C., 16, 117, 138

E

Eichinger, B., 19, 38–39, 41, 42, 79, 121
evaluation of impact, 141–143, *142*
    talent analytics using development strategies matrix in, 143–151, *144–150*
evidence-based management (EBM), 13

F

feedback-rich environments, 80–82, *81*
Financial Times, 125
Flaum, J. P., 95
Frame it, establishing and connecting the business case with, 90–94, **91–93**
Frame it, See it, Own it, Connect it formula, 88–89, *89,* **90**
Freedman, A., 119
Fuller, T., 73

G

Garr, S., 22
Gerstner, L., Jr., 73
Gettysburg battle, 34
Goff, M., 46
Google, 125
growth mindset, 16–17, 95
growth potential, 37–38
    aspiration in, 42–43
    capability in, 42
    development strategies matrix and, 58–60, **58–60**
    finding the best assessment tool or method for, **45,** 45–47

learning agility in, 19, 38–39
versus performance, 54

H

habit vs. experimentation, 112–114, *114*
Hagemann, B., 48
Handbook of NeuroLeadership, 82
Harvard Business Review, 88, 124, 138
Hoffer, E., 37
Huselid, M. A., 61

I

I4CP, 9
IBM, 73
identification of essential skills, 94–105, **100–104**
identifying essential skills through, 94–105, **100–104**
Impact Maps, 14, *14,* 14–15, 91, **91–93,** 94, 141, 143; *see also* evaluation of impact
    personal business case, 94
implementation intentions, 85
Improving Performance: Managing the White Spaces on the Organization Chart, 40
Ingvar, D., 85
Institute for Corporate Productivity, 46
Intentional Development, 6, 7, 15–16, 20, 42, 64
    being behavior specific in, 117–118
    build it in; don't bolt it on for, 74–78, **76–77**
    career development vs. development for a position and, 120
    career transitions and, *119,* 119–120
    challenge of acknowledging our weaknesses in, 111
    competency flow in, 120–121
    Connect it, leveraging new skills through, *118,* 118–122, *119,* 136
    creating a cadence of development for, 80
    creating feedback-rich environment for, 80–82, *81*
    curating experiences in, 121–122
    development essentials in, 71–86

development targets in, 99–100, **100–101**
focusing on the critical few competencies, 73–74, *74*
Frame it, establishing and connecting the business case through, 90–94, **91–93**
Frame it, See it, Own it, Connect it formula for, 88–89, *89*, **90**
habit vs. experimentation in, 112–114, *114*
having a planned and targeted impact for development initiatives, 72–73
introduction to process of, 87
learning map and, 101–105, **102–104**
learning theory and, *67*, 67–71
making development "sticky" for, 82–85, *84*, **85**
mass personalization in, 78
matching the development strategy to the talent for, 82
never learning alone in, 85–86
owning it all and changing it all in, 111–112
owning our successes in, 109–110
Own it, building skills through, 105–118, **107**, **109**, *110*, *114*
plan for, 79, 106–108, **107**
reflection, role of, 105, 108–109, 118, 168, 169, 171
seeing small victories in, 115–117
seeing success in others in, 114–115
tracker for, 109, **109**, *110*
understanding there's more than one path to development for, 78–79
Intentional Leadership Coaching, 123–125
applied in the right situations, 129–130
coachable employees and, 137–140
defining, 126–127
focused on the right players, 127–129, *128*
integrated, 132–133
keys to effective, 127–133, *128*
reimagining the approach to, 133–140
rights skills for delivering, 130–132
uncoachable employees and, 134–135

J

journey, Strategy-Driven Leadership Development (SDLD), *153*, 153–155
derailers in, 157–165
SDLD maturity and, 155–157, *156*

K

Key Developmental Experiences (KDEs), 118–119
Key Leadership Competencies (KLCs), 118–119
Knowles, M., 69, 72

L

Lake, R., 30
leadership brand, 33
leadership development, *see* Intentional Development; Strategy-Driven Leadership Development (SDLD)
business case for, 8, *8*
corporate spending on, 4
current status of, 3–10, *7*, *8*
fairy tale, 1–3, 167–172
where and how to invest in, 36–37
Leadership Machine, The, 37
learning agility, 19, 38–39
learning map, 101–105, **102–104**
learning theory and Intentional Development, *67*, 67–71
leveraging of new skills, *118*, 118–122, *119*
lock in the learning, 89, 108, 109
Loew, L., 22
Lombardi, V., 123
Lombardo, M., 19, 38–39, 41, 42, 79, 121

M

Making Change Happen One Person at a Time: Assessing Change Capacity Within Your Organization, 49
managerial courage, 126
mantras, 32, 84–85, 101–104, 107, 109, 142, 171
mass personalization, 78
matching the development strategy to the talent, 82

Mattone, J., 48
maturity, Strategy-Driven Leadership
  Development (SDLD), 155–157, *156*
McKinsey Company, 8, 91
McRaven, W. H., 116–117
Meister, J., 124
mental agility, 41
mentoring, 123–124
  peer, 136
Microsoft, 39
mindset, growth, 16–17, 95
mission-critical skills, 31
modeling
  of the future, not the past, 28
  of roles and processes, not jobs and
    departments, 29
  of top performance, 27–28
  up-to-date, 33
model of development, 70-20-10, 81,
  86, 126
Mooney, T., 13–15, 91, 141
Mount, M. K., 46
multi-rater assessments, 46–47
multi-rater feedback (360s), 98–99, 121

N

Nadella, S., 39
network analysis, 43–44
networks, social, 43, 86
NeuroLeadership Institute, 16, 32
neuroscience, 15–16, 49, 75, 84, 131
New Talent Management Network, 43
Now Discover Your Strengths, 19

O

online learning, 67
*Organizational Capability: Competing
  from the Inside Out*, 30
organizational culture, 8, 72
organizational demand, *21*, 21–23, 97
organizational strategy and, 23–24
  strategy-driven competencies and, 24–33
organizational development (OD)
  professionals, 3, 6
organizational strategy, 23–24
organization charts, 40–41
Own it, building skills through, 105–118,
  107, 109, *110, 114*

P

Paths to Improvement, 79
peering, 124
peer mentorship, 136
people agility, 41
People Capability Maturity Model, 155–156
People Innovation Lab, 125
Pepermans, R., 39
performance effectiveness over time,
  43–44
  development strategies matrix and,
    58–60, **58–60**
  finding the best assessment tool or
    method for, **45**, 45–47
  versus growth potential, 54
Performance Management Research, 125
performance reviews, 68
personal business case, 94
Peterson, D., 71
pivotal roles, 18–19, 63
Project Oxygen, 125
promotability, 37–38
Prusak, L., 49

R

Ramstad, P., 18, 19
Ready, D., 88
Resilience Advantage, The, 78–79
results agility, 41
Retooling HR, 21
Rock, D., 16, 83, 86
"rule of three," 32
Rummler, G., 40
Russell, C. J., 9

S

Schein, E., 8
Schenck, J. xi
Scullen, S. E., 46
SDLD, *see* Strategy-Driven Leadership
  Development (SDLD)
See it, identifying essential skills through,
  94–105, **100–104**
self-awareness, 41
self-paced learning, 67
self-serving bias, 111
simplicity in competencies, 32–33

single manager assessments, 45–46
Southwest Airlines, 32–33
stand-and-deliver training, 67
strategy-driven competencies, 17, 20, 24–25
   aligning and integrating all your talent
      processes, 29–30, *30*
   building your leadership brand, 33
   coming alive for every employee,
      31–32
   defined, 21
   identifying and validating, 25–33
   kept simple, 32–33
   modeling roles and processes, not jobs
      and departments, 29
   modeling the future, not the past, 28
   modeling top-performance, 27–28
   not simply copying others' work, 28–29
   not started from scratch, 26–27
   s up-to-date models of, 33
   tied to development, 32
   validating, 30–31
Strategy-Driven Leadership Development
   (SDLD), 5–10, *7*
   assessing organization and leadership
      capability and (*see* assessment
      of organization and leadership
      capability)
   coaching in (*see* Intentional Leadership
      Coaching)
   competencies in (*see* strategy-driven
      competencies)
   evaluation of impact of (*see* evaluation
      of impact)
   foundations of, 13–20
   growth mindset in, 16–17
   Impact Maps for, *14*, 14–15
   journey to (*see* journey, SDLD)
   organizational demand and (*see*
      organizational demand)
   pivotal roles in, 18–19
   strengths movement and, 19–20
   talent assessment in, 18
Strategy-Driven Leadership Development
   Maturity Curve, 20
Strengths Movement, 19–20
Swisher, V., 41

T

Talent Reviews, 48–49
   designing and implementing, *50*,
      50–57
testing or executive assessments, 47
tools for identifying and validating
      strategy-driven leadership
      competencies, 25–33

U

Ulrich, D., 30

V

validating competencies, 30–31
Vantilborgh, T., 39

W

Weiss, A., xi, xvi
Wellman, N., 138
Whitehurst, J., 138
Willyerd, K., 124

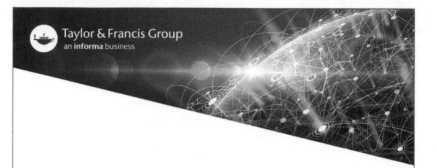